The MACH-10 PM
AI-Powered Product Management at
Hypersonic Speed
Jason M. Riggs

Hypersonic Publishing

The MACH-10 PM: AI-Powered Product Management at Hypersonic Speed™

Published by Hypersonic Publishing

Encinitas, California, USA

https://mach10pm.com

ISBN: 979-8-9940323-1-2 (Paperback)

First Edition

Dedication

To my wife Megan, and our boys, Riley and Logan, thank you for your brutal honesty, your love, and your patience through every obsessive late-night writing session and AI rabbit hole.

To my brother Scott, who's always had my back.

To my small circle of friends, who lend me an ear far more often than I probably deserve.

To my mentors and colleagues, past and present, who've challenged me to think bigger, lead better, and stay curious.

To my failures, big and small, loud and quiet, public and private. You've been my harshest critics and my greatest teachers. I wouldn't be here without you.

And to the next generation of product leaders: may you build with purpose, lead with empathy, and always stay just a little uncomfortable. That's where the magic happens.

Foreword: The MACH-10 PM

Leading at MACH-10

I've seen PMs burn months chasing a perfect roadmap, only to watch the market shift overnight. That's when you realize: speed without clarity is chaos, and clarity without speed is irrelevant.

When I started in product management over twenty years ago, success came down to grit, intuition, and listening. We lived in customer feedback, debated on whiteboards, sat through user interviews, and aligned teams through will, charm, or compromise. There was no playbook, just iteration and a belief that great products come from being close to the problem and obsessed with the customer.

But the game has changed.

Product management isn't just instinct and influence anymore. Today's PMs have something new: intelligence at scale. Not just data, but insight. Not just tools, but collaborators, systems that think, analyze, and recommend. With AI, we're no longer buried in decisions or stretched thin. We can move faster and lead with a level of clarity we never had before.

PMs are at a crossroads. You can resist the shift, or lean in early and become the kind of leader this era demands: adaptive, strategic, and unreasonably effective.

This is your field guide. A playbook for scaling your thinking, outpacing competitors, and leading with precision and creativity at hypersonic speed.

AI is one of the most powerful tools we've ever had. But it only works if you use it.

Let's dive in.

Jason Riggs

Encinitas, California

Contents

Contents

Introduction: Build for Speed, Lead with Precision

Why MACH-10 Product Management Is the Future

MACH-10.

That's over 7,600 miles per hour, ten times the speed of sound. Fast enough to cross the U.S. in under 20 minutes.

No human has flown that fast. Even NASA's X-43A, an unmanned marvel, peaked at Mach 9.6. At MACH-10, everything changes: friction becomes fire, and feedback must be instant. One wrong move and you're out of the sky.

That's product management now.

It used to be about process: gather requirements, validate assumptions, plan, build, ship. You were rewarded for follow-through.

That world is gone.

Now you're expected to see trends early, make calls with half the data, and lead through uncertainty, fast.

You're not just managing features. You're navigating chaos.

This book is for PMs who are done reacting.
For builders who want leverage, not just output.
For leaders chasing clarity and speed.

Inside are frameworks and prompts to help you operate at this altitude. To guide without delay. To use AI as a real advantage, not a gimmick.

You don't need to work harder. You need to work **with precision**.

The future won't slow down.
Let's get you to MACH-10.

Chapter 1: The MACH-10 PM in the Age of AI

When AI Stops Being Optional

Every era of technology reaches an inflection point. It's a moment when the pace accelerates, the rules shift, and what once defined "best practice" becomes the new constraint. For product managers, that moment is now. Artificial intelligence, particularly generative AI, has moved beyond the domain of researchers and early adopters. It's now deeply embedded in the tools, workflows, and decisions that define the craft of product management.

Just as the internet reshaped commerce and mobile revolutionized communication, AI is redefining how we envision, build, and scale products. The difference? This wave is moving faster than anything before it. AI isn't just a technology shift; it's an organizational, strategic, and personal transformation. It's altering how teams operate, how strategies are formed, and how competitive advantages are won.

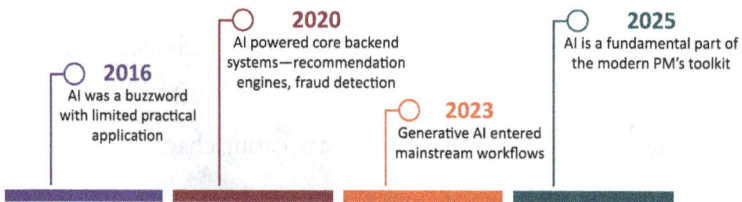

2016
AI was a buzzword with limited practical application

2020
AI powered core backend systems—recommendation engines, fraud detection

2023
Generative AI entered mainstream workflows

2025
AI is a fundamental part of the modern PM's toolkit

The role of a product manager has always been dynamic, but the rise of AI has raised expectations. The MACH-10 PM is now expected to deliver sharper insights, make data-informed decisions faster, lead leaner teams, and harness AI to drive measurable business outcomes. Those who master AI won't just work faster: they'll think differently, unlocking a level of strategic leverage that changes the very definition of the role.

AI is not here to replace product managers. But the ones who lean in and master it will move faster and think sharper than the ones who don't. This is not about squeezing out more productivity. It is about multiplying your leverage.

> **AI won't replace PMs. But PMs who embrace AI will outpace those who don't**

The Machine's Job vs. Your Judgment

AI is not a silver bullet. It excels at certain tasks but has very real limitations that demand human oversight, judgment, and creativity. Knowing where to lean on AI — and where to lead it — is critical to unlocking its true value, especially in handling tasks that overwhelm humans: spotting patterns in massive datasets, distilling thousands of survey responses into a handful of themes, generating quick ideas to spark exploration, and projecting likely outcomes based on historical data.

But it stumbles on the nuance of understanding the politics inside your org, weighing ethical trade-offs, or connecting dots across industries in ways that spark true innovation. That's where you come in.

The best PMs don't outsource their thinking to AI. They collaborate with it.

The MACH-10 PM Advantage

Being AI-powered, or more precisely AI-augmented, means weaving AI into every stage of the product lifecycle. It is not about doing the same work faster. It is about transforming the work itself. AI raises the ceiling of what is possible and lowers the cost of exploration.

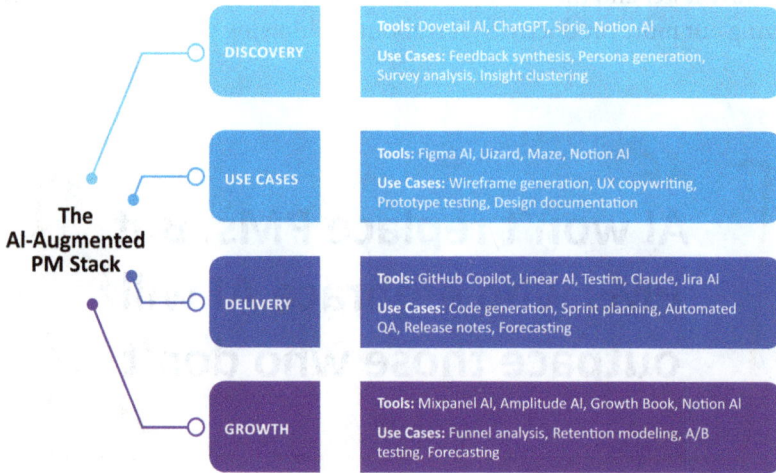

The AI-Augmented PM Stack

DISCOVERY	**Tools:** Dovetail AI, ChatGPT, Sprig, Notion AI **Use Cases:** Feedback synthesis, Persona generation, Survey analysis, Insight clustering
USE CASES	**Tools:** Figma AI, Uizard, Maze, Notion AI **Use Cases:** Wireframe generation, UX copywriting, Prototype testing, Design documentation
DELIVERY	**Tools:** GitHub Copilot, Linear AI, Testim, Claude, Jira AI **Use Cases:** Code generation, Sprint planning, Automated QA, Release notes, Forecasting
GROWTH	**Tools:** Mixpanel AI, Amplitude AI, Growth Book, Notion AI **Use Cases:** Funnel analysis, Retention modeling, A/B testing, Forecasting

AI turns speed into clarity and clarity into leverage

MACH-10 PMs integrate AI seamlessly into their workflows to not only move faster, but also think more clearly. Here are just a few ways they gain an edge:

- Execute at high velocity without sacrificing quality or due diligence.

- Drive alignment across teams by sharing AI-enhanced insights in real time.

- Unlock new opportunities by simulating scenarios and market shifts before they happen.

The payoff is not just speed. It is clarity. PMs can pivot more decisively, pursue bigger bets, and ensure that strategic choices are backed by both data and insight.

The Tools That Make This Real

This book is not about AI theory. It is about application. AI is now as essential to your toolkit as your roadmap, backlog, or prioritization framework. The MACH-10 PM applies AI intentionally, selecting tools and workflows that create measurable value.

As shown below, here are five AI tools to try this week and three ready-to-use prompts that can deliver ROI immediately. These are not theoretical suggestions.

They are battle-tested entry points to start making AI a part of your operating rhythm:

5 Tools to Try This Week

ChatGPT	For summarizing research and generating ideas
Notion	For writing product briefs and meeting notes
Figma	For early design exploration
Jira	For backlog insights
Amplitude	For product usage analytics

3 Prompts to Test Today

1. What are 3 ways AI can improve my current product workflow?
2. Summarize the key benefits of AI for PMs in under 100 words.
3. List 5 AI tools that align with my product goals.

The Wake-up Call

AI isn't optional. It's the foundation of competitive advantage and the core skillset for modern product leaders. The PMs who rise fastest will weave it into every decision, sprint, and roadmap while preserving the human touch that builds trust and creativity.

Ask yourself: Where could AI amplify your impact? What could you streamline or reinvent if you treated it as a collaborator instead of a background tool?

At MACH-10 speed, survival isn't about effort. It's about clarity.

Mindset Shift to Adopt

If AI is the engine, the PM is the driver. Adopting it isn't about adding a tool but rethinking how you create value and lead.

Reframe your role in three ways:

- From execution to leverage — let AI multiply impact.

- From gathering to generating — let AI cut noise so you find the signal.

- From manager to leader — treat AI as a partner in outcomes.

When embraced, AI becomes a creative ally. It frees you to focus on vision, opportunity, and decisive action. The MACH-10 PM gains the edge by combining human judgment with machine precision.

Reflective Questions

Use these questions to identify immediate opportunities for AI integration:

- Which responsibilities could AI meaningfully improve?

- How will you preserve creativity while leveraging AI speed?

Exercise:

Run a one-week AI-augmented sprint, and apply what you have learned with this practical exercise:

- Pick one product decision or discovery task.

- Use at least two AI tools to analyze, ideate, or forecast.

- Compare the output and document what AI improved, missed, and what to refine.

Key Takeaways

- MACH-10 PMs shift from process to strategy.

- AI handles patterns and predictions; humans add nuance.

- AI fluency is now a competitive requirement that gives PMs clarity and confidence.

Chapter 2: Stop Doing, Start Multiplying

You're Not Measured by Output, but Impact

Every PM knows the feeling of being stretched too thin: a backlog full of requests, meetings eating the calendar, and customer interviews you never quite have time to schedule.

AI multiplies your impact. It scales you across discovery, prioritization, design, and delivery, clearing repetitive work so you can focus on shaping decisions that truly drive the product's trajectory. What matters now isn't how much you produce, but how effectively you steer.

The traditional PM skill set still matters: defining vision, prioritizing initiatives, synthesizing insights, driving collaboration, and measuring outcomes. Those foundations don't disappear; they evolve. With AI taking on execution-heavy tasks, your value shifts from producing deliverables to guiding the insights and strategies that determine success.

The Old PM Playbook: All Friction, No Leverage

Without AI, the core skills of product management were a manual, high-friction process that often looked like this:

- **Defining vision:** Endless workshops and opinion-driven debates instead of clear data.

- **Prioritizing impact:** Stalled in subjective meetings where the loudest voice often won.

- **Synthesizing research:** Weeks spent coding interview notes and survey data.

- **Driving collaboration:** Teams misaligned due to scattered, outdated information.

- **Measuring outcomes:** Decisions made from lagging, quarterly reports.

The Traditional PM Skill Set

Core product management skills remain as vital as ever. PMs are still expected to:

Define and communicate a compelling product vision

Prioritize initiatives based on user and business impact

Conduct research and synthesize user insights

Collaborate with design, engineering, and go-to-market teams

Measure outcomes and iterate with discipline

AI shifts PMs from tasks to impact

The New Skill Stack: Your AI-Era Arsenal

The MACH-10 PM builds on those fundamentals with new capabilities built for an AI-powered environment. Intelligent systems aren't just assisting anymore; they're active collaborators in the product lifecycle. Your job is to guide, integrate, and elevate that intelligence with human judgment.

New skills for the AI-era PM:

- **Build AI literacy:** Know where AI shines, where it fails, and how to use it responsibly.

- **Practice prompt engineering:** Craft inputs that generate useful, actionable outputs.

- **Strengthen data fluency:** Validate and challenge AI analysis with your own judgment.

- **Apply ethical judgment:** Anticipate risks and ensure automation doesn't create blind spots.

But as AI takes on more executional load, the relative weight of each skill is shifting. PMs must complement foundational competencies with a new set of capabilities designed for an AI-augmented environment.

The New PM Skill Stack

To lead effectively in the AI era, PMs must develop fluency in several emerging domains:

Skill	Description	Why It Matters Now
AI Literacy	Understanding how AI works and where it applies	Enables smarter tool use and better collaboration with data teams
Prompt Crafting	Writing effective prompts to extract value from AI tools	Turns AI into a creative and analytical partner
Data Fluency	Interpreting and questioning AI-generated insights	Prevents blind trust in outputs
Ethical Judgment	Navigating bias, privacy, and fairness in AI decisions	Builds trust and avoids reputational risk
System Thinking	Designing workflows that integrate AI seamlessly	Ensures AI enhances, not disrupts, team dynamics

Multiplication beats grind: scale your impact, not your hours.

Some PMs are already making this shift, moving from theory into practice. They're showing what AI-driven product management looks like by reshaping workflows, accelerating decisions, and raising the bar for high-velocity product leadership.

Stop Doing, Start Multiplying

AI shifts a PM's value from execution to impact. What matters is shaping decisions that define the product's path.

- Set bold visions competitors can't follow.

- Spot hidden advantages in crowded markets.

- Bet on trends before they're obvious.

How AI-Driven PMs Work

In one enterprise hardware team, onboarding time dropped by half after the PM prototyped flows with AI overnight. That's MACH-10 leverage: turning clarity into speed, and speed into results.

Tactics of AI-Driven PMs:

- **Design AI-native experiences** — making intelligence part of the user journey, not an add-on.

- **Prototype at speed** — moving from idea to feedback in hours, not weeks.

- **Stress-test with simulation** — modeling user behavior to surface risks early.

- **Inspire new thinking** — pushing teams past the limits of old playbooks.

> **" The best AI-augmented PMs aren't coders—they're translators. They connect business goals, user needs, and AI capabilities "**

Making the Leap

AI can feel overwhelming. Some PMs fear replacement, others stall on where to start. The truth: it multiplies those who embrace it and leaves behind those who don't.

To adapt, focus on:

- **Staying curious** — explore new AI capabilities and keep learning.

- **Sharing what you learn** — help your team climb the curve faster.

- **Focusing on outcomes** — tie experiments to business results.

- **Building durable skills** — strengthen AI literacy, prompt mastery, and strategic judgment that evolve as the tools do.

The Rise of the MACH-10 PM

The PM of the future won't be measured by tasks completed. They'll be known for multiplying impact, making sharper bets, and steering products at a pace competitors can't match. No longer just facilitators, they'll be integrators of intelligence — using AI to amplify judgment, expand influence, and lead with greater impact. Your impact isn't in tasks done, but in leverage created.

5 Tools to Try This Week	
Dovetail	For synthesizing user interviews
productboard	For prioritization and feedback clustering
miro	For brainstorming and mapping
Linear	For sprint planning
slack	For summarizing team discussions

3 Prompts to Test Today
1. What are the top 3 skills I need to develop as an AI-augmented PM?
2. How can I use AI to shift from execution to leverage?
3. What tasks in my week could be delegated to AI?

Mindset Shift to Adopt

You are no longer just a facilitator. You are an integrator of intelligence. Your value comes from leverage: sharper insight, tighter priorities, and a clear product vision. The PM who thrives in the AI era learns quickly, trusts their judgment, and turns tools into lasting impact.

To step into this role:

- Start with intent. Use AI where it moves the needle, not where it creates noise.

- Focus on outcomes. Tie experiments to real business results.

- Build judgment, not dependence. Know when to trust AI, when to challenge it, and when to bet on your instincts.

Reflective Questions

- Where are you spending time on tasks that don't require your judgment?

- If you had five extra hours each week, where would you invest them?

- Which decisions on your plate truly need sharper insight, not just faster execution?

Exercise

Choose two recurring tasks you handle manually, such as backlog grooming or summarizing research. Run them once the old way and once with AI. Share the results with a teammate. Don't just compare the outputs, notice how the conversation shifts when AI is in the room.

Key Takeaways

- AI sharpens the fundamentals. It doesn't replace product management — it strengthens the core.

- Use it to elevate judgment, spending less time on noise and more on decisions that matter.

- Expand your influence with clearer insights, faster alignment, and stronger stakeholder narratives.

- Lead with greater impact by pairing machine speed with human creativity, trust, and vision.

Chapter 3: Out-Leverage, Don't Out-Work

AI as a Catalyst for Strategic Leverage

A I isn't just about speed. It expands the boundaries of product management, reshaping how you think, what you prioritize, and the scale of your impact. The best PMs don't use AI to check more boxes. They use it to ask sharper questions, place bolder bets, and steer with greater confidence.

That's the multiplier effect: cognitive, operational, and market leverage that turns the same resources into outsized results. You're no longer trying to outwork complexity. You're out-leveraging it, multiplying the impact of every decision you make.

AI doesn't replace judgment. It refines it. It clears the noise, surfaces hidden options, and lets you run more scenarios in less time. The payoff is moving from gut feel to evidence, from reaction to foresight, and from incremental progress to accelerated advantage.

At Qualcomm, discovery could drag for months. By the time a customer interview summary hit my desk, the market had already shifted. At GoPro, I saw the flip side—a new streaming feature surfaced in discovery and within weeks we had prototypes in customer hands. The speed of that loop was the difference between relevance and irrelevance.

The Three Levers of AI Power

In product management, leverage is the ability to generate outsized results from the same resources. AI adds three new forms of leverage that compound when used together:

- **Cognitive leverage**—process more information, spot patterns faster, and model outcomes before committing resources.

- **Operational leverage**—clear bottlenecks like backlog grooming, synthesis, and release notes so you spend energy where it matters.

- **Market leverage**—adapt faster than competitors with predictive personalization, real-time optimization, and adaptive systems that stand out in crowded markets.

Used with purpose, these levers create momentum—decisions get clearer, teams move faster, and outcomes scale with precision.

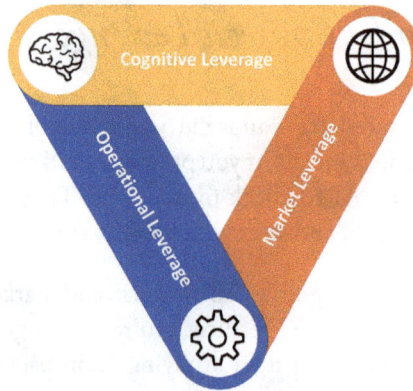

Leverage multiplies impact across thinking, operations, and markets

The triangle illustrates a simple but powerful truth: leverage in product management is multi-dimensional. You can't outwork complexity by sheer effort, but you can out-leverage it by multiplying the impact of every decision you make.

Cognitive leverage elevates how you think and connect dots others can't see. Operational leverage expands your capacity by removing friction from repetitive, low-value work. Market leverage extends your reach, allowing your product to adapt and differentiate faster than competitors.

Individually, each dimension is powerful. Together, they create compounding effects—the kind of acceleration that turns a good product leader into a force multiplier for the entire organization. The point isn't to choose one; it's to orchestrate all three in balance, so your mental bandwidth, team operations, and market impact reinforce one another.

From Leverage to Application

Understanding the three forms of leverage is only half the battle. The real test is knowing where to apply them in your day-to-day product work. Not every use case deserves equal attention. Scattershot AI adoption creates noise instead of impact. Strategic PMs focus on the moments where leverage compounds, where a single insight, automation, or market signal can ripple through the system and unlock outsized results.

It's not about more AI. It's about sharper placement. The best leverage points live inside friction, delay, and stalled decisions. Smart PMs map the slowdowns, then strike. Applied with precision, leverage doesn't just create speed. It shifts trajectory.

Area	Strategic AI Use Case
Customer Insight	Analyze thousands of support tickets or reviews to surface unmet needs
Product Strategy	Simulate market scenarios or competitor moves using LLMs
Roadmapping	Generate and evaluate roadmap options based on constraints and goals
Team Alignment	Use AI to summarize meetings, decisions, and trade-offs for stakeholders
Experimentation	Rapidly prototype and test ideas with AI-generated assets and flows

The smartest PMs don't use AI everywhere. They use it where it multiplies impact

Turning Leverage into a Living Strategy

For years, strategy moved at the pace of headcount. Collecting signals, framing problems, and testing solutions stretched into months. By the time you had answers, the market had already shifted.

AI changes that rhythm. Now you can run a strategy loop in days—or even hours:

- **Sense** what's happening—scan user behavior, market signals, and product data in real time.

- **Frame** the problem—look at it from multiple angles so you don't lock in too early.

- **Model** scenarios—test how different choices play out before you burn resources.

- **Decide** with confidence—make the call backed by data, not just instinct.

- **Learn** fast—feed those lessons back into the next cycle.

The power isn't in running the loop once. It's in running it more often than your competitors. When they're still reacting to last quarter's insights, you're already shaping the next move.

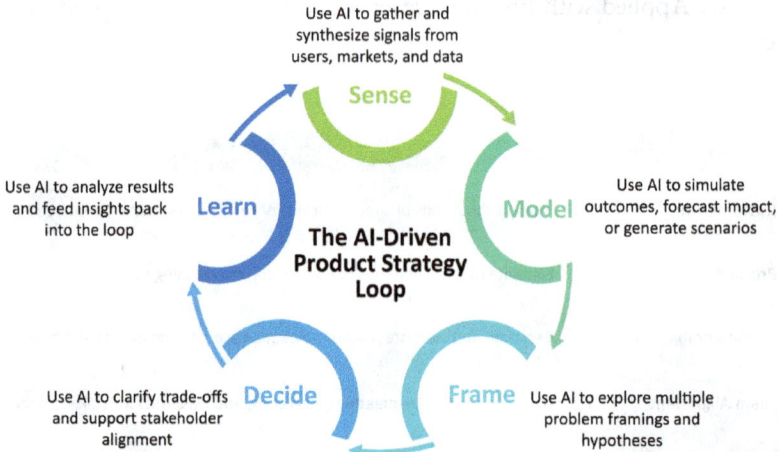

Sense → Frame → Model → Decide → Learn. The faster you loop, the further you pull ahead

Asking the Right Strategic Questions

AI is only as sharp as the questions you throw at it. Treat it like a search bar and you'll get generic answers. Push it with pointed, ambitious questions and you'll uncover opportunities others miss.

Examples:

- *"What unmet needs are buried in the last 1,000 support tickets?"* → surfaces pains before churn shows up in metrics.

- *"If we ship Feature X, how are our top three competitors most likely to respond?"* → helps you play offense, not just defense.

- *"What monetization models would feel natural to users but are rare in our category?"* → spots growth paths hidden in plain sight.

- *"Looking at past launches, what risks are we underestimating?"* → prevents blind spots from turning into expensive mistakes.

Great PMs don't just react to inputs; they frame better questions. AI multiplies the payoff of curiosity when you use it this way.

From Reactive to Proactive

AI helps you step off the hamster wheel. With continuous signals and scenario modeling, you can spot early warning signs in user behavior, stress-test concepts before committing resources, and cut through opinion wars with data-backed clarity. The payoff is more time steering the future instead of cleaning up the past.

Signals from the Edge

At a SaaS company I worked with, a PM began by evangelizing AI's value through quick demos that cleaned up their messy backlog. Once the team saw results, she shifted to educating them, running short sessions on how to write effective prompts. Soon, they were experimenting with workflows, using AI to generate prototypes overnight. Within a quarter, she had elevated the conversation with leadership, framing AI not as a tool but as a strategic driver. What began as small wins turned into lasting cultural change, just as the pyramid below illustrates. Start with visible wins, teach the craft, and scale the system until it becomes how the team works.

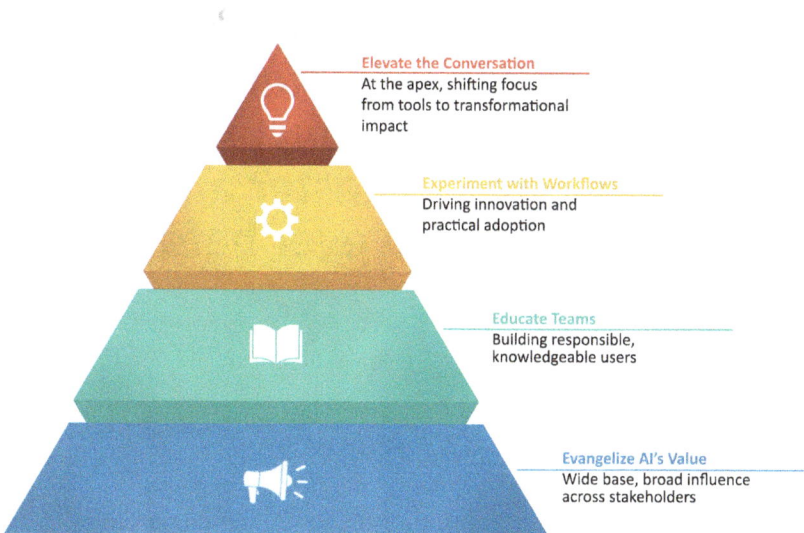

Elevate the Conversation
At the apex, shifting focus from tools to transformational impact

Experiment with Workflows
Driving innovation and practical adoption

Educate Teams
Building responsible, knowledgeable users

Evangelize AI's Value
Wide base, broad influence across stakeholders

Lead the charge. Inspire change. Transform with AI

Leading with AI

The PMs who break through aren't the ones chasing every shiny tool—they're the ones weaving AI into how their teams operate. Leading with AI is more cultural than technical.

It looks like this:

- Rallying teams around AI's upside instead of letting fear dominate.

- Teaching peers how to use it responsibly and effectively.

- Running small but real experiments in your workflows and showing the lift.

- Shifting the conversation from *"Which tool should we try?"* to *"How can AI help us solve this problem better?"*

Once you lead this way, AI stops being a side hustle and becomes part of your product DNA.

> **" AI isn't just a tool; it's the strategic amplifier that transforms product vision into reality faster and smarter than ever before "**

Orchestrate, Don't Execute

Bringing AI into product management isn't about bolting on tools. It's about changing how you think. You shift from tasks to systems and from outputs to outcomes.

Modern product managers need to:

- Design for adaptability across the system instead of optimizing in silos.

- Create loops of continuous learning instead of chasing delivery alone.

- Use AI as a lens that expands your reach instead of a time-saver.

- Orchestrate a stack that compounds instead of hoarding tools.

The PMs who thrive here aren't just faster at execution. They lead with a broader view, one that turns AI into a source of leverage, foresight, and lasting competitive edge. They connect signal to action and near term to strategy so momentum compounds. In product, leverage beats hustle every time.

5 Tools to Try This Week	
✳ Claude	For strategic prompt writing
mixpanel	For predictive analytics
Figma	For rapid prototyping
productboard	For roadmap generation
ChatGPT	For scenario modeling

3 Prompts to Test Today

1. What are 3 strategic bets we could make using AI insights?

2. Simulate 2 roadmap scenarios based on different constraints.

3. What are the top risks we're underestimating this quarter?

Mindset Shift to Adopt

In the era of AI-driven product management, your role is no longer about keeping projects moving—it is about amplifying intelligence and shaping outcomes. You're no longer just coordinating work. You're orchestrating intelligence. Your value comes from the leverage you create, not the tasks you complete.

To step into that role:

- Value clarity over volume—information is cheap, insight drives action.

- Prioritize the decisions where your judgment changes the outcome.

- Treat AI like a teammate, not a vending machine.

- Focus on ensuring the right work gets done, not just more work.

Reflective Questions

- What are three strategic product questions you wish you could answer more effectively today?

- How might integrating AI into your decision-making process change your priorities and resource allocation?

Exercise

Select one major product decision from the past quarter. Use an AI tool to model at least two alternative scenarios for that decision. Compare the projected outcomes and document the insights you wish you had before making the original decision.

Key Takeaways

- AI strengthens strategic foresight—helping PMs detect trends, analyze competitive landscapes, and model market shifts.

- Scenario modeling with AI mitigates risk—testing multiple paths before committing resources.

- AI enhances alignment—turning abstract visions into data-backed narratives that resonate with stakeholders.

- The PM's role evolves—from solely intuition-driven leadership to a hybrid model blending instinct and AI-derived insight.

Chapter 4: Insight at Speed

Transforming Research with Machine Intelligence

In Chapter 3, we saw how AI acts as an amplifier—broadening your thinking, streamlining operations, and sharpening your market edge. Discovery is often where this leverage first becomes visible. It's the moment product insight takes root, and the quality of your questions defines the opportunities you uncover.

Traditionally, discovery has been the heartbeat of product management: the place where insights form, ideas emerge, and direction begins. But it has also been slow, biased, and limited by capacity. Interviews, research, and analysis take time, and in fast-moving markets, that delay can cost you your lead.

AI changes the tempo. With the right tools and prompts, you can scan markets, uncover unmet needs, create realistic personas, and validate ideas in a fraction of the time. It's not just efficiency—it's perspective. Human intuition, paired with machine-driven insight, makes discovery faster, deeper, and far more actionable.

Where Discovery Breaks Down

For decades, discovery followed a predictable path:

- Scheduling and conducting user interviews

- Reviewing support tickets and NPS responses

- Analyzing competitor features and pricing

- Running surveys or small-scale experiments

That process worked, but only up to a point. Insights depended on how many interviews you could run, how much feedback you had time to code, and whether the team acted on it. Sample sizes were small. Interpretations were fuzzy. Execution was hit or miss.

As a result, many teams struggled to run discovery with the rigor and consistency it required. Key decisions were often made with partial or outdated information—leaving blind spots that slowed progress or steered products off course.

Discovery at MACH-10

AI surfaces hidden patterns that manual research misses, turning discovery from a bottleneck into a strategic advantage. AI enables you to:

- Cast a wider net — analyze massive volumes of user, competitor, and market data in seconds.

- Spot patterns fast — surface recurring themes, gaps, and anomalies with precision.

- Validate at scale — test concepts broadly before committing resources.

- Predict with confidence — anticipate shifts in user behavior or market dynamics before they happen.

AI doesn't replace discovery—it amplifies it. It makes the process continuous, proactive, and data-rich, giving you clearer insights and creating a learning engine that compounds over time to keep you ahead of shifting markets and evolving customer needs.

TASK	TRADITIONAL	AI-AUGMENTED
Market Research	Manual reading of reports	Summarize trends from dozens of sources instantly
Persona Creation	Based on interviews and assumptions	Generated from behavioral and demographic data
Idea Generation	Brainstorming sessions	Prompt-based ideation with LLMs
Feedback Analysis	Manual tagging and clustering	NLP-powered sentiment and theme extraction
Competitive Analysis	Feature-by-feature comparison	AI-generated SWOT and positioning maps

Discovery is a rhythm: ask sharper, learn faster, refine endlessly

Tools That Turn Noise Into Signal

When applied with intent, AI elevates every stage of discovery. The goal isn't to adopt every tool but to match the right capability to the right step.

Natural Language Processing (NLP): Clusters and summarizes thousands of feedback points from tickets, surveys, and social channels, giving PMs an evidence-based view of user sentiment.

Example: A Fintech PM finds "account linking delays" in 40% of complaints after running 18 months of support logs through NLP.

Generative Models: Build personas and simulate conversations so PMs can test messaging, onboarding flows, and feature ideas before writing code.

Example: A Healthtech PM tests onboarding chats with three simulated patient personas to see which explanations resonate.

Predictive Analytics: Surfaces early market signals before they appear in analyst reports, helping PMs anticipate shifts and outpace competitors.

Example: An enterprise SaaS PM spots demand for "eco-certified packaging" six months before rivals and launches early.

Collaborative AI Platforms: Merge inputs from design, engineering, marketing, and sales into a shared insight hub, reducing misalignment and building consensus faster.

Example: A SaaS PM uses an AI workspace to combine mockups, objections, and feedback, speeding prioritization.

These tools turn discovery into a constant pulse, keeping PMs ahead of markets and users.

Tool	Use Case
/ Claude	Market synthesis, idea generation, persona drafting
delve.ai	Auto-generated personas from web and CRM data
Crayon / Kompyte	Competitive intelligence and battlecards
Typeform + AI	Smart survey analysis
Dovetail + AI	Thematic analysis of user interviews

AI tools turn discovery into a constant pulse

Prompting for Discovery

The real unlock in AI-powered discovery isn't just the tools — it's the prompts. A prompt is how you aim the system. Well-framed prompts surface sharper insights, expose blind spots, and open possibilities you might otherwise miss. Poorly framed ones return generic noise.

Strong prompts create clarity. They refine your thinking, challenge assumptions, and help you see angles you wouldn't otherwise consider. Once you learn to steer AI effectively, it stops being a shortcut and becomes a genuine partner — one that works at scale, on demand, and without fatigue.

Here's what this looks like in practice:

- **Market Trends**
  Prompt: *"Summarize the top five emerging trends in the B2B SaaS collaboration space based on recent news and reports."*
  Result: A PM at a productivity startup spots a spike in asynchronous video adoption. It wasn't on their roadmap, but it reshapes Q3 priorities and puts them ahead of competitors.

- **Persona Generation**
  Prompt: *"Create three user personas for a mobile budgeting app targeting Gen Z professionals."*
  Result: The team learns many users juggle multiple side gigs. That insight drives a new feature focused on flexible, multi-income budgeting.

- **Idea Validation**
  Prompt: *"What are the potential risks and points of differentiation for launching a voice-based productivity assistant in 2025?"*
  Result: The team discovers user interest is high, but privacy concerns dominate. They design an MVP that prioritizes transparency and user control over voice data.

- **Problem Framing**
  Prompt: *"Given this user feedback, what are three possible root problems we should explore?"*
  Result: A health app learns churn isn't about missing features, but about too many push notifications. They redesign notification logic and reduce drop-off.

Prompting isn't a hack—it's a core PM skill. Learn to ask better questions, and you'll unlock better insights. Ask the right ones consistently, and you'll not only move faster, you'll move smarter.

The AI-Powered Discovery Loop

Discovery has always been central to product management, but also one of its slowest steps. Interviews, surveys, and market scans often take weeks, and by the time answers arrive, the market has already shifted. AI compresses that cycle into hours or days, letting you move with the market instead of chasing it. Speed isn't the only gain. You also widen your lens, catching insights that traditional methods miss.

The loop works best as a rhythm, not a one-off exercise:

Ask — Begin with a sharp, specific question. Instead of "What do users want?" try: "What unmet needs appear in our last 1,000 support tickets?"

Generate — Use AI to cluster themes, surface hypotheses, or highlight options you might not have considered.

Validate — Pressure-test those insights quickly with real user data, lightweight experiments, or scenario modeling.

Refine — Feed results back into the system, tightening the next round of questions and increasing fidelity.

Each cycle builds momentum. Instead of waiting weeks for static research, teams operate in a continuous loop of learning, testing, and adjusting. Insight compounds. Direction gets clearer. And execution becomes smarter with every turn.

Example:
A PM investigating trial-user churn feeds AI a batch of feedback logs. The model clusters responses around "setup complexity." Instead of launching another survey cycle, the PM prototypes a simplified onboarding flow, tests it with real users, and validates the fix within days.

The loop itself is simple: ask sharper questions, learn faster, and refine continuously. Over time, the impact compounds. Teams that adopt this rhythm stop reacting to change and begin anticipating it, building confidence and momentum with every cycle.

Scan the landscape,
identify patterns,
generate hypotheses

Explore

Refine
Adjust direction based
on findings and team
alignment

**AI-Powered
Discovery Loop**

Frame
Recast the problem
from multiple
angles

Validate
Test assumptions with
lightweight simulations
or feedback synthesis

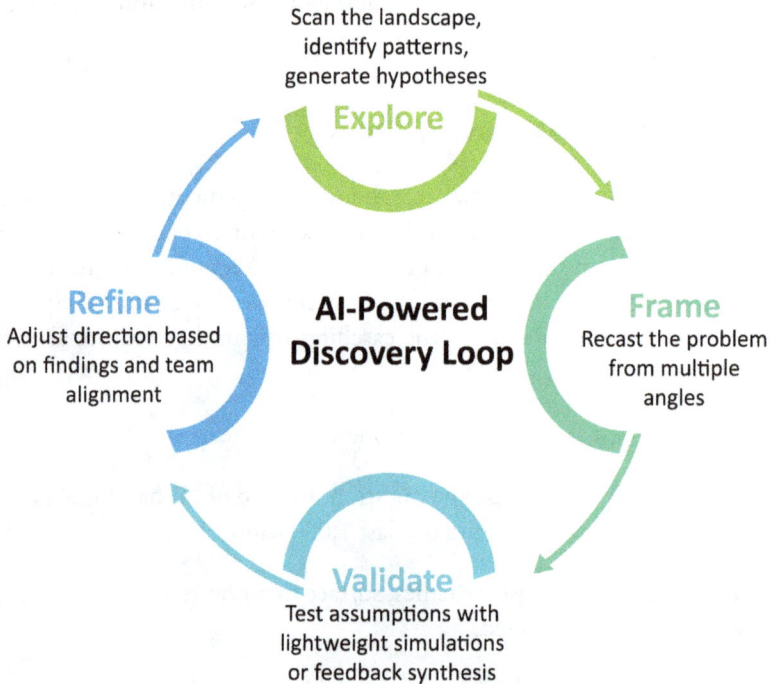

Sharpen questions, learn faster, refine constantly

Great discovery doesn't come from luck or genius. It comes from building systems that learn faster than the market changes. When you integrate AI into that system, you stop chasing insights and start compounding them. The more you feed the loop, the more elevated your intuition becomes—and the further ahead you can see.

Landmines and Blind Spots

AI can accelerate discovery, but acceleration without discipline is risky. Without care, you may move faster in the wrong direction or base decisions on faulty inputs. Common pitfalls include:

- **Hallucinated insights** — AI can produce outputs that sound entirely plausible but are factually incorrect. Without verification, these errors can slip into your decision-making and lead to misguided strategies. Always validate AI-generated findings with trusted data or human expertise before acting on them.

- **Bias amplification** — AI models reflect the data they are trained on. If that data contains biases, the outputs will reinforce them. This can skew your understanding of user needs, underrepresent certain groups, or perpetuate harmful stereotypes. Proactively diversify your data sources and include bias checks in your workflow.

- **Over-reliance** — AI can process information faster than you can, but it cannot replace human empathy, ethical judgment, or contextual understanding. Over-reliance risks turning decision-making into a mechanical process that ignores nuance and the "why" behind user behavior.

Treat AI like a microscope, not a compass. It can magnify what you see, but only you can decide where to look and what to do with the insight. The combination of AI's analytical power and your strategic discernment is what produces consistently high-quality outcomes. When you direct its power with focus and intent, AI becomes less of a tool and more of a trusted partner in driving smarter, faster decisions.

> " **AI expands the surface area of discovery—letting you explore more ideas, faster, with less bias** "

Discovery as a Weapon

In dynamic markets, the speed and depth of discovery often define whether you lead or lag. AI helps you explore more ideas with the same team, validate faster, and align stakeholders around a shared, data-informed view. If you've ever said, "We don't have time for discovery," AI eliminates that excuse by making it a continuous habit across the organization.

The edge doesn't go to the team with more data — it goes to the team that acts faster on it.

5 Tools to Try This Week	
Dovetail	For clustering feedback
ChatGPT	For persona generation
Sprig	For survey analysis
Figma	For early design exploration
Notion	For insight synthesis

3 Prompts to Test Today
1. Generate 3 user personas for our target segment.
2. What are the top unmet needs in our support tickets?
3. Reframe this user problem from 3 different angles.

Mindset Shift to Adopt

AI broadens your discovery surface, exposing opportunities you might miss manually. The real shift is in how you think and act as a product leader. Explore more ideas quickly and let speed strengthen your analysis rather than weaken it. Diversify inputs by pulling from broad data sources such as social sentiment, market reports, and competitor reviews. Start with AI, but always validate insights through interviews or experiments. Discovery becomes a continuous habit powered by AI's reach and your judgment.

Reflective Questions

- Where in your discovery process do you most often hit bottlenecks?

- What insights are you missing because your team is moving too slowly?

- If discovery were continuous instead of episodic, how would your roadmap look different?

Exercise

Pick a current product question. Run a mini-discovery sprint using AI: cluster recent customer feedback, draft personas, and generate hypotheses. Compare what you learn in a few hours against what you normally get in weeks.

Key Takeaways

- Discovery is no longer bound by time or bandwidth.

- AI turns it into a continuous loop that compounds intuition.

- The best PMs don't run more interviews, they learn faster and act sooner.

- The edge goes to the team that moves first, not the one that gathers the most data.

Chapter 5: Roadmapping at MACH-10

From Static Plans to Adaptive Strategy

R oadmaps are supposed to align teams. Too often, they drain momentum instead. The problem is not intent, it is how slow and brittle the process becomes once politics and stale data creep in.

I once sat in a roadmap review where a PM spent 20 minutes defending Feature X for Q2. The data was stale and the plan unraveled. Another PM brought three AI scenarios tied to trade-offs. Instead of debating the past, the team debated the future and unlocked momentum. The conversation shifted from defending decisions to testing them, and the tone of the room changed instantly.

AI makes that shift possible. Roadmaps adapt as conditions change. Decisions get clearer, trade-offs explicit, and alignment faster. RICE, Kano, and Weighted Scoring still matter, but AI turns them into living systems that learn as you do. You move from gut calls to signals you can defend, and from static planning to a responsive model that evolves alongside your product.

Why Roadmaps Fail and How AI Rewrites the Rules

Traditional roadmapping blends science and politics. Incomplete data and competing agendas create rigid plans that go stale fast.

AI adds clarity with:

- **Forecasting:** Simulate effects before committing.

- **Scenario testing:** Compare growth, retention, or tech-debt paths.

- **Real-time updates:** Adjust often without losing alignment.

- **Assumption tracking:** Make bets explicit and testable.

Roadmapping shifts from defense to exploration. With AI, it becomes a living model that adapts on signal, helping teams avoid the noise, hidden trade-offs, and rigidity that cause roadmaps to fail.

The Classic Roadmap Headaches

If you've built a roadmap by hand (and defended it to a skeptical exec team), you've faced these:

| Subjective prioritization driven by gut feel or the loudest voice |

| Limited visibility into effort, risk, and dependency impact |

| Rigid formats that fall apart the moment priorities shift |

| Stakeholder misalignment that leads to churn and frustration |

Roadmap headaches: gut feel, low visibility, rigid formats, misalignment

Now flip the lens. The next chart shows how AI turns roadmapping from manual judgment into modeled decisions. It surfaces trade-offs early, tests assumptions, and keeps plans aligned as signals change.

How AI Levels Up Roadmapping

AI doesn't remove tough calls—but it gives you a sharper lens to make them. It enables:

Skill	Traditional	AI-Augmented
Idea Scoring	Manual evaluation	AI-based scoring using historical data and impact models
Dependency Mapping	Engineering-led	AI-assisted visualization of technical and team dependencies
Scenario Planning	Spreadsheet modeling	AI-generated roadmap variants based on constraints
Stakeholder Alignment	Meetings and decks	AI-generated summaries and trade-off analyses

AI makes roadmaps smarter, clearer, and stronger

This shift goes beyond mechanics. It changes who decides and what gets priority. With AI, dependency mapping, scenario planning, and alignment scale faster. Daily work like grooming, analytics, and feedback takes less time, freeing PMs to focus on higher-value decisions. It tilts the balance from firefighting to forward motion, turning time once lost to coordination into space for strategy.

Meetings stop being status readouts and become decision engines. Instead of chasing inputs, PMs can finally focus on outcomes. This is where the real leverage shows up when AI clears the runway so you can actually fly.

The Efficiency Shift: Operational Work Before and After AI

Operational work consumes a huge share of PM time. AI keeps ownership intact but removes the drag of repetition, creating space for better decisions. Here's a side-by-side look at how workflows shift with AI. From backlog grooming to analytics and customer feedback, these changes unlock speed and reduce friction.

They also strip out the hidden coordination tax that quietly bleeds teams dry. Status updates stop clogging calendars because the data stays live and current. Instead of chasing artifacts, PMs can interrogate the system directly and get instant clarity. The feedback loop tightens, the noise floor drops, and signals finally stand out across the org. This is the difference between nudging progress and unleashing it.

OPERATIONAL TASK	TRADITIONAL APPROACH	AI-AUGMENTED APPROACH
Backlog Grooming	Manual review of tickets, estimates, and dependencies	AI suggests ticket prioritization based on effort, impact, and velocity
Sprint Planning	Manual selection and scoping based on team memory or gut instinct	AI clusters backlog items into sprint-ready groups based on past patterns
Meeting Summaries	Handwritten or post-meeting recap emails	AI auto-generates summaries in real-time with action items
Stakeholder Updates	Manual status reports in slides or email	AI creates digestible snapshots from Jira, Slack, and analytics tools
Analytics & Reporting	Manual dashboard creation and filtering	AI highlights anomalies, trends, and suggests questions to explore
Customer Feedback Analysis	Time-consuming tagging and synthesis	AI identifies themes across surveys, tickets, and reviews automatically

Snapshots fade. Living roadmaps win

AI Turns Roadmaps Into Living Systems

AI upgrades static frameworks into adaptive systems that move as fast as the market:

- **Scenario modeling:** Test multiple paths in minutes and surface trade-offs early.

- **AI scoring:** Rank features with data, not gut feel, cutting through bias and politics.

- **Predictive dashboards:** Spot shifts early so you can pivot before problems hit.

- **Dynamic resourcing:** Rebalance teams quickly around the highest-leverage bets.

- **Assumption tracking:** Tag risky bets and auto-check if they still hold.

The goal isn't to replace judgment but to amplify it with sharper inputs, faster cycles, and a clearer view of what's ahead. The roadmap stops fossilizing and starts evolving with the business. Teams stop defending guesses and start steering with live signals, shifting roadmaps from static promises into active decision engines.

Turning roadmaps into living systems takes more than theory. It takes the right tools wired into your daily workflows. The platforms below do more than track work. They turn static slides into adaptive engines that learn and adjust in real time. They transform planning from a quarterly ritual into a competitive advantage.

When live data flows back into your roadmap, decisions stop lagging behind reality and start shaping it in the moment. Strategy stays in sync with execution, even as the ground shifts under your feet.

When roadmaps behave like living systems, momentum stops being fragile and starts becoming inevitable.

> **"**
> ## AI doesn't make decisions for you—it makes the decision space clearer
> **"**

From Concept to Execution

Knowing what makes a roadmap dynamic is only half the job. The real shift happens when you wire those principles into your daily stack. AI tools give you the scaffolding to do it at speed, pulling live data into planning and turning static slides into living systems.

The table below shows where to start. Each tool aligns to a core roadmapping skill, helping you build a system that learns, adapts, and accelerates with your product.

Tools That Support AI-Driven Roadmapping

Skill	Use Case
Aha! + AI	Roadmap generation and prioritization suggestions
Productboard + AI	Feature scoring and feedback clustering
Craft.io	Scenario planning and roadmap visualization
ChatGPT / Claude	Prompt-based roadmap ideation and trade-off analysis
Jira + AI plugins	Sprint forecasting and backlog grooming

AI tools turn roadmaps into living systems

Wiring Roadmaps for Real-Time Speed

Living roadmaps are not theory. They are built with the right tools wired into your daily workflows. The platforms below merge AI with your stack to turn static slides into adaptive engines that adjust in real time.

When live data flows back into your roadmap, decisions stop lagging behind reality and start shaping it in the moment. Strategy stays in sync with execution, patterns surface faster, and roadmaps stop being artifacts and start acting like operating systems.

Dynamic Roadmaps

Instead of a linear timeline, think of your roadmap as a responsive model—something that can be interrogated, reshaped, and optimized as new information surfaces.

Input: Voice of customer, product vision, business goals

Simulate: Model multiple paths and risks

Prioritize: Balance short- and long-term gains

Align: Communicate scenarios with transparency

Adapt: Revisit monthly—not quarterly

Strategy beats snapshots. Roadmaps must evolve.

Tools alone are not enough. Sharper inputs mean nothing if the roadmap stays static. To make it a living system, you have to guide it with precision, and that starts with the prompts you feed into it.

Precision Prompts for Smarter Bets

As with discovery, the value of AI in roadmapping comes from the prompts you feed it. Sharp prompts cut through noise and give you outputs you can trust and act on. They frame decisions with clarity, surface trade-offs early, and anchor the conversation in data instead of opinion.

- **Feature Scoring:** "Score these 10 feature ideas based on user impact, technical effort, and strategic alignment."

- **Scenario Planning:** "Generate three roadmap scenarios for Q4: one for growth, one for retention, one for technical debt."

- **Trade-Off Analysis:** "Compare the pros and cons of prioritizing Feature A over Feature B given current capacity and user feedback."

- **Stakeholder Summary:** "Summarize the trade-offs and rationale behind our Q3 roadmap for executive review."

These prompts expand the decision space so you can guide the room instead of reacting to it. They shift the conversation from defending ideas to pressure-testing them.

Once the right prompts sharpen your decisions, the next step is building a roadmap that can evolve with them.

Roadmaps That Think and Evolve

Static roadmaps are relics. An AI-enhanced roadmap is a responsive model that can be interrogated, reshaped, and optimized as new information surfaces.

The Dynamic Roadmap Flow:

- **Simulate:** Model paths, risks, and dependencies.

- **Prioritize:** Balance short- and long-term outcomes.

- **Adapt:** Revisit often to keep the roadmap relevant.

This turns the roadmap into a unifying signal that drives clarity, confidence, and commitment across the org.

AI-Augmented Roadmap Workflow

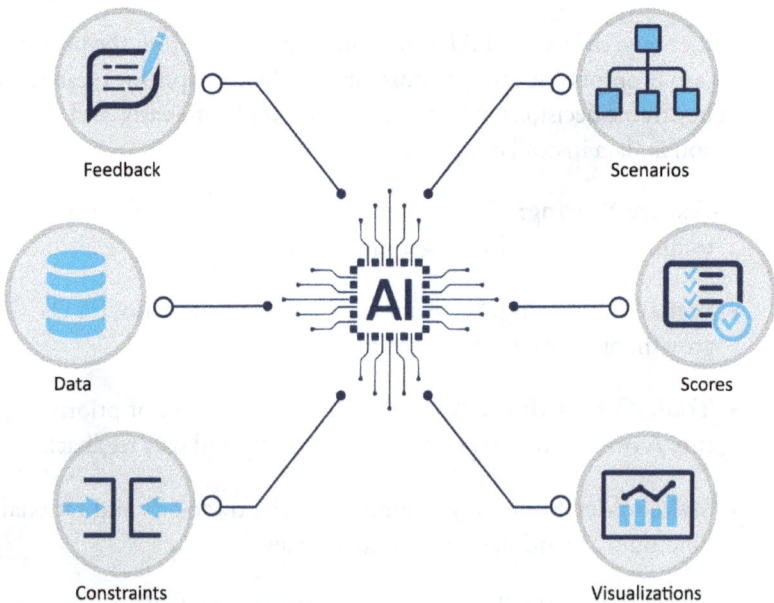

Feedback   Scenarios   Data   Scores   Constraints   Visualizations

Roadmaps that evolve as fast as your market.

Frameworks That Don't Freeze

Frameworks once helped PMs manage priorities, but they were rigid snapshots that went stale fast. AI makes them adaptive, evidence-driven, and tied to live signals instead of gut feel.

- **RICE with AI:** Traditional RICE relied on guesses. AI uses adoption data and predictive models so prioritization is grounded in evidence, not debate.

- **Kano with AI:** Kano once needed surveys and manual coding. AI processes thousands of comments to reveal which features delight, which are expected, and which frustrate.

- **Weighted Scoring with AI:** Old scoring was static and politicized. AI makes weights dynamic so decisions follow strategy, not influence.

Don't memorize frameworks. Keep them flexible so prioritization reflects reality. A roadmap is not a list of promises. It is a bet on the future.

⚠️ Landmines and Blind Spots

AI sharpens prioritization, but only with discipline. Speed without control creates noise, not clarity.

The traps:

- Chasing easy metrics over lasting impact

- Skipping context and judgment in the rush to automate

- Losing trust when an "AI-picked roadmap" lacks clear rationale

The fix: Pair AI with product insight, keep your data clean, and always explain the why.

The PM who masters this balance earns both speed and credibility. They stop arguing for their roadmap and start commanding the room. They become the person leadership trusts to move fast without losing the plot. They turn prioritization from a debate into a decision engine, where trade-offs are surfaced early and choices are made with confidence. That is how AI becomes leverage instead of noise.

Roadmapping as a Strategic Ritual

With AI in the loop, roadmapping shifts from defense to strategy. Teams model scenarios, clarify trade-offs, and test assumptions before burning resources. A roadmap isn't a defense exhibit but a compass. It should guide decisions in real time, not just validate old ones. Build it to explore, not justify, and the PM who keeps it evolving earns trust and the confidence to place bigger bets. The best roadmaps don't lock you in, they keep you moving, adjusting course as the market shifts under your feet.

5 Tools to Try This Week

productboard	For feature scoring
ChatGPT	For trade-off analysis
Jira	For effort estimation
miro	For roadmap visualization
Claude	For stakeholder summaries

3 Prompts to Test Today

1. Score these 10 features based on impact and effort.

2. Generate 3 roadmap scenarios for Q4.

3. Summarize the trade-offs in our current roadmap.

Mindset Shift to Adopt

AI will not make your decisions, but it will make the options clearer. Use it as a spotlight for possibilities and risks, not a substitute for leadership. Treat every AI output as a starting point, not an answer. Think of your roadmap as evolving, use scenarios to drive trade-off conversations, and balance qualitative insight with AI's quantitative rigor to ground strategy in both data and vision. The goal is not to follow what the system says, but to lead with sharper judgment because of it.

Reflective Questions

- Where do stakeholders in your org most often stall alignment?

- What trade-offs do you routinely struggle to make visible?

- How might scenario modeling change the conversation around your roadmap?

Exercise

Take your current roadmap and model two alternate versions using AI — one focused on growth, another on retention. Share them with your leadership team and watch how the debate shifts when the trade-offs are explicit.

Key Takeaways

- Your roadmap isn't a courtroom exhibit. It's your compass.

- Build it to explore futures, not defend the past.

- Scenarios are rehearsal, not distraction.

- Alignment grows from clarity, not persuasion.

- The PM who treats the roadmap as a living system earns trust, momentum, and the right to place bolder bets.

Chapter 6: Beyond Guesswork

How AI Turns Feedback into Advantage

E very PM has lived through it: a launch derailed by a bad bet. I once worked with a team that shipped a "must-have" feature based purely on gut feel. Two sprints later, usage was flat, churn ticked up, and the roadmap had to be ripped apart. The rework wasn't just expensive — it crushed morale. The problem wasn't a lack of effort. It was guessing.

AI takes guessing off the table. It doesn't replace judgment, but it gives you a constant stream of sharper signals so you can decide with confidence. Instead of waiting weeks for research cycles or debating whose spreadsheet is right, you can tap into feedback in real time, spot patterns instantly, and act while the market is still listening. PMs who learn to harness those signals will move faster, make better calls, and build products that stay ahead of the curve.

From Concept to Prototype in Hours

The path from idea to prototype used to be measured in weeks. Sketching, wireframing, and building clickable flows was slow, resource-heavy work that often stalled momentum. With AI, that timeline collapses. You can generate wireframes, explore multiple design directions, and map user flows in hours instead of weeks. The blank page is no longer a barrier. Designers start with strong, AI-assisted drafts and focus their energy on refining the details. The work moves faster, collaboration tightens, and judgment sharpens because teams are iterating on something tangible almost immediately.

Where Research Slows You Down

Old-school research is slow, narrow, and noisy. Interviews and surveys take weeks, the loudest voices dominate, and data lives in silos. By the time findings surface, they're stale. PMs end up making decisions on outdated signals while the market moves on.

How AI Rewrites the Playbook

AI doesn't just speed research. It rewrites the playbook. Old methods waited for quarterly reports and polished decks. With AI, thousands of comments cluster in minutes. Anomalies surface before they become fires. Shifts in sentiment show up while you can still act. Patterns appear early, giving PMs an edge while others wait for slides.

With AI, you don't wait for reports. You get continuous radar. Signals flow daily, so your roadmap reflects what is happening, not what already happened. The shift is simple: reactive reports become proactive clarity. Think lagging reports versus real-time radar.

How AI Rewrites the Playbook

AI enables real-time, large-scale, and multi-channel research. Here's how:

TASK	TRADITIONAL	AI-AUGMENTED
Interview Analysis	Manual Note-Taking and Tagging	AI-Generated Summaries and Sentiment Analysis
Survey Analysis	Manual Charting and Filtering	NLP-Based Theme Extraction and Clustering
Feedback Monitoring	Periodic Reviews	Continuous Monitoring Across Channels
Insight Synthesis	Human-Led Pattern Recognition	AI-Assisted Insight Generation and Prioritization

Guessing dies when research runs in real time

How AI Enables Smarter, Scalable Research

Collecting feedback is easy. Making sense of it is the grind. AI turns haystacks into needles:

- **Clustering & Topic Modeling** — thousands of comments distilled into themes.

- **Sentiment & Intent Detection** — separating emotional rants from real requests.

- **Persona Refresh** — keeping goals and blockers current with live data.

- **Trend Spotting** — seeing shifts right after launches, not quarters later.

Now feedback fuels growth instead of bogging you down.

AI isn't theory anymore. The tools below turn these capabilities into daily practice, giving PMs leverage where it matters most.

Tools that Elevate Research

Tool	Use Case
Dovetail + `AI	Interview transcription, tagging, and insight generation
EnjoyHQ	Centralized research repository with AI-powered search
Sprig	In-product surveys with AI analysis
thematic	AI-driven feedback analysis from multiple sources
/ Claude	Prompt-based synthesis and persona refinement

Radar that never sleeps keeps you ahead of reality

Prompts That Unlock Insight

Even with the best tools, the value of AI comes down to the questions you ask. Strong prompts act like precision instruments: they tell the system what to focus on, how to structure the output, and what kind of clarity you need.

Here are a few high-value prompt patterns that turn raw data into insight you can use:

Interview Summary
Prompt: "Summarize key pains and requests from this transcript. Include direct quotes."
Why it matters: Saves hours and preserves the customer's voice.

Another prompt: "Highlight the top three unmet needs and suggest one product opportunity for each."
Why it matters: Surfaces problems and actionable paths forward.

Theme Extraction
Prompt: "From 500 survey responses, list the top five themes with counts."
Why it matters: Turns piles of feedback into a prioritized shortlist.

Another prompt: "Cluster 1,000 support tickets by product area and rank by urgency."
Why it matters: Helps you act on what matters most.

Sentiment & Drivers
Prompt: "Classify app-store reviews by sentiment and top complaint drivers."
Why it matters: Goes beyond positive/negative to show the "why."

Another prompt: "Analyze social mentions, tag sentiment, and link issues to product features."
Why it matters: Adds market-wide context beyond your own channels.

Persona Refinement
Prompt: "Update Persona A's goals, fears, and success metrics based on this dataset."
Why it matters: Keeps personas fresh instead of static.

Another prompt: "Use churn survey results to refresh Persona B's motivations and blockers."
Why it matters: Grounds personas in live behavior.

The goal isn't to replace your judgment—it's to accelerate the grunt work so you can apply sharper thinking where it counts. With prompt libraries like these, you bring consistency to your analysis and rigor to your decision-making.

Radar That Never Sleeps

While others guess from stale data, you stay ahead. AI builds an always-on listening loop that:

• Streams feedback from tickets, reviews, NPS, and social.

• Pushes AI-summarized insights to squads through weekly digests.

• Flags sentiment drops or theme spikes right after releases.

Quarterly research looks neat in a slide deck but delivers stale insights. AI creates an always-on radar: streaming signals, surfacing risks early, and aligning teams around the same truth. The real advantage isn't speed alone — it's a culture that moves from debating data to acting on it.

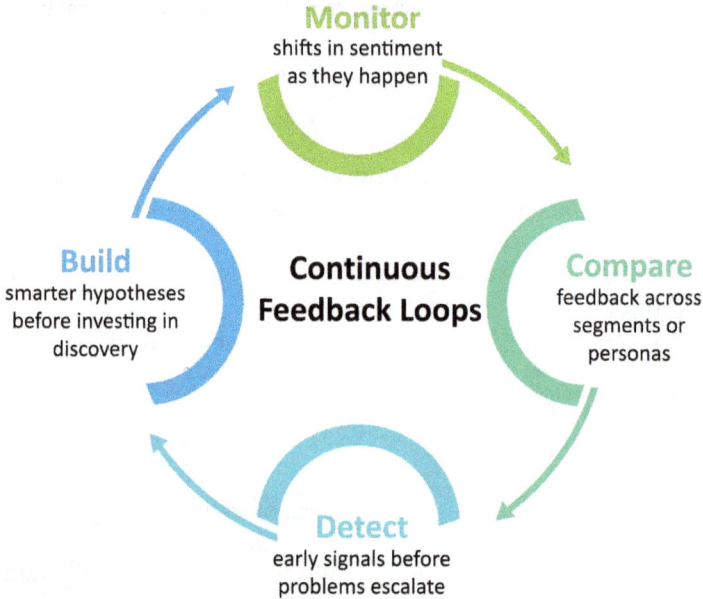

Signals matter more than hunches. AI keeps you ahead

Monitor, compare, detect, and build — a feedback loop that keeps your product aligned with reality.

When this loop is running, feedback becomes a shared resource across the organization. No one is caught off guard in a roadmap meeting because everyone sees the same signals in near real time.

From Data to Decisions

Feedback without action is just noise. AI makes it easier to turn signal into movement. It helps you close the loop faster, with less guesswork and more clarity.

Use it to:
Prioritize with precision
Don't just stack tickets. Rank issues by size of pain and impact on real users. Know what's worth fixing first.

Mirror customer language
Use their words, not yours. When copy sounds like your users, it lands faster and sticks longer.

Spot underserved segments
Find patterns buried deep in the data. See the customers no one's building for—yet.

Prevent churn early
Catch red flags before they turn into lost accounts. Stay one step ahead of regret.

Signals from the Edge

A SaaS PM watched NPS tank. The signals were in support tickets, but buried under noise and delay. By the time reports surfaced, churn was already underway. With AI, those patterns appeared in hours instead of weeks — clarity that only mattered if the team could move on it.

⚠ Landmines and Blind Spots

AI is a microscope — powerful, but it needs the right guardrails:

- **Sentiment tunnel vision** — don't confuse tone with root cause.

- **Lost nuance** — side comments and contradictions still need human ears.

- **Privacy pitfalls** — mishandled feedback can erode trust.

AI gives speed; your job is to apply judgment.

Research as a Competitive Advantage

Everyone has data. What separates the leaders is how quickly they turn signals into action. AI-powered listening lets you see shifts early, validate assumptions faster, and move before competitors even notice the trend. While others debate stale insights, you're already shipping the next advantage.

Another Signal from the Edge

A retail tech startup used AI to scan reviews across its category, not just its own product. While rivals clung to surveys, they spotted "checkout speed" as a top frustration. Acting fast, they launched a one-click flow months ahead of competitors, lifting conversions and winning a reputation as the most user-friendly option. What looked like luck was disciplined listening, applied faster.

You don't need a custom AI stack. A few tools and sharp prompts can surface insights that shift strategy and execution. Here are some to try this week.

Guessing is obsolete. Clarity wins.

5 Tools to Try This Week

Dovetail	For tagging and clustering feedback
Sprig	For survey analysis and sentiment tracking
ChatGPT	For summarizing interviews and reviews
Amplitude	For behavioral insights
Notion	For synthesizing research notes

3 Prompts to Test Today

1. Summarize the top 5 complaints from our last 500 support tickets.

2. What are the most common themes in our app store reviews?

3. How has user sentiment shifted over the last 3 months?

Lock in the Loop

Turn AI-powered listening into a habit, not a one-off. Pick your top three high-signal channels such as support tickets, app reviews, or sales call notes, run them through an AI clustering pass, and share the top three actionable themes with your team. Aim for one quick win, one mid-term fix, and one learning opportunity each week. When everyone sees the same signals, alignment gets easier. End each cycle by asking, *"What changed because of what we learned?"*

Mindset Shift to Adopt

You're not short on feedback. You're short on capacity to process it.

Use AI to clear the noise, sharpen the signal, and put your judgment where it counts.

Reflective Questions

- Where does your feedback loop slow or distort the truth?

- If you had reliable AI summaries every week, what would you change first?

Exercise

Pick a recent batch of support tickets or reviews, run them through an AI analysis, and identify your top three actionable insights. Share them with your team and decide one action to ship within two weeks.

Key Takeaways

- AI enables continuous, multi-channel listening and fast synthesis that turns noise into decisions.

- Automated clustering, sentiment, and trend detection reveal patterns that human-only workflows often miss.

- Prompt libraries standardize summaries, themes, and persona updates for repeatable rigor.

- Always-on feedback loops reduce risk and improve responsiveness after each release.

- Guardrails around nuance and privacy keep insights trustworthy and compliant.

Chapter 7: Prototype at MACH-10

From Concept to Clickable in Hours

I 've seen teams wait three weeks just to get an onboarding flow mocked up. By the time the first prototype landed in Figma, the market had shifted and competitors were ahead. The PM was frustrated, the designer was buried, and the sprint was stuck in a holding pattern.

I worked with a PM who waited weeks for user feedback. By the time insights landed, the market had already moved. AI closed that loop, pulling patterns from live usage so the team could respond in days, not months.

That's the power of collapsing a three-week cycle into 72 hours. AI didn't replace the designer. It cleared the runway for focused execution.

I've also seen the opposite play out—teams using AI to cut weeks into hours. Instead of waiting for a single polished prototype, they spin up multiple variations in a day, test quickly, and learn which direction deserves real investment. The market keeps moving, but now they're moving with it.

Why Design Becomes the Bottleneck

By the time a project reaches design, the clock is already ticking. Market windows are short, engineering timelines are fixed, and pressure to "just ship" often overshadows exploration. Even great design teams hit the same roadblocks:

- **Slow iteration** — only a few variations get tested because each one takes too much manual effort.

- **Misaligned intent** — briefs and outputs drift, creating rework.

- **Manual prototyping** — building and updating high-fidelity flows takes too long.

- **Copy delays** — waiting on UX writers stalls velocity.

These frictions narrow the solution space, frustrate teams, and force compromises on quality or scope just to keep schedules intact.

The good news: this is where AI can clear the runway.

How AI Accelerates Design

AI doesn't replace designers, it clears the runway for them. The tools handle the heavy lifting so humans focus on higher-value thinking. Rapid wireframing, draft copy, flow mapping, and real-time critique close the gap between idea and execution.

The payoff is more ideas explored, more paths tested, and stronger confidence before the first line of code. Instead of burning weeks for a handful of options, teams can spin up variations in hours and decide which ones deserve investment. With more iterations on the table, blind spots shrink and insights surface earlier.

AI shifts design from a slow sequence of handoffs to a fast cycle of exploration. Bottlenecks turn into traction, and that momentum compounds into sharper decisions and faster feedback.

The real advantage is simple: design stops dragging and starts driving.

How AI Accelerates Design

TASK	TRADITIONAL	AI-AUGMENTED
Wireframing	Manual Sketching or Figma Work	AI-Generated Wireframes from Text Prompts
UX Copywriting	Written by PMs or Designers	AI-Generated Microcopy, Tooltips, and CTAS
Flow Mapping	Whiteboard Sessions	AI-Generated user Flows Based on Goals
Usability Testing	Manual Recruitment and Testing	AI-Simulated user Interactions and Feedback

Design at full throttle: speed and insight unlocked by AI

AI Tools That Unblock Design

The PM–designer toolkit is more than Figma files and sticky notes. AI adds tools that boost creativity and cut friction. They don't replace design craft, they clear space for it.

- **Wireframe generators:** Turn prompts or sketches into layouts in minutes.

- **UX writing AIs:** Draft on-brand microcopy for flows, CTAs, and error states.

- **Journey mapping platforms:** Build flow diagrams from goals or feature lists.

- **Accessibility checkers:** Flag compliance risks before inclusivity becomes a scramble.

Together, these tools expand options, shorten cycles, and let PMs contribute earlier without draining design resources.

Tools that Elevate Prototyping

Tool	Use Case
Figma +	Generate design variations, auto-fill copy, create instant wireframes
uizard	Turn sketches or text prompts into clickable prototypes
Framer	Build responsive website prototypes from natural language prompts
maze + `AI	Rapid usability testing with AI-generated user flows and analytics
Canva + `AI	Quick mockups, simple UI concepts, and lightweight prototyping for marketing experiences

AI lights the afterburners on prototyping

Designing at MACH-10

AI can be a fast, creative partner, but only if you give it the right instructions. The quality of your prompts dictates the quality of the output. Vague inputs deliver vague results. Clear, specific prompts give AI a real chance to produce something useful.

Think of prompts like briefing a designer. The more context you provide — audience, tone, constraints, goals — the stronger the first draft. The clearer the brief, the more the AI feels like a teammate instead of a tool.

Examples:

- **Wireframe Generation:** "Design a mobile onboarding flow with goal selection, personalization, and a progress dashboard."

- **UX Copywriting:** "Draft concise, friendly microcopy for password reset, including error and confirmation states."

- **Flow Mapping:** "Map a freemium sign-up path from entry through upgrade to paid subscription."

- **Design Critique:** "Review this screen for clarity, hierarchy, and accessibility; list three improvements."

Clear prompting shifts design from slow iteration to fast exploration. Instead of waiting weeks for mockups and approvals, teams move from idea to prototype in hours.

From Static to Dynamic Prototyping

A SaaS team once spent four weeks on onboarding — research, mockups, copy, review. With AI wireframes and copy, they tested three variations in a week. By the next sprint, real user feedback drove decisions. The result: a 20% lift in trial-to-paid conversion, and onboarding flipped from bottleneck to advantage. That's the AI shift in prototyping: ideas become interactive and testable almost as fast as you describe them. Instead of waiting weeks for design bandwidth, you're experimenting in real time. And the faster you test, the faster you learn. Teams that embrace this loop turn uncertainty into momentum.

With AI in the loop, teams can:

- **Generate multiple design variations instantly** — giving PMs and designers clear options to compare and test before committing.

- **Test flows with simulated users early** — surfacing usability issues before engineering gets involved and reducing costly rework.

- **Align stakeholders faster** — sharing polished prototypes early so conversations focus on decisions and trade-offs, not guesswork.

These aren't just efficiency wins. They are structural advantages.

Key Improvements with AI:

When AI becomes part of the design loop, the payoff shows up in five big ways:

- **Speed:** Concepts move from idea to testable artifact in hours, not weeks.

- **Confidence:** Real-time signals replace assumptions.

- **Breadth:** Teams explore more options before narrowing down.

- **Quality:** Automated checks catch usability and accessibility issues early.

- **Collaboration:** PMs and designers stay aligned by contributing directly to iteration.

The value is not just faster delivery, but learning sooner, sharpening smarter, and building with greater certainty. The teams that win are the ones that turn speed into sustained advantage by turning every iteration into insight. The graphic below highlights the key improvements teams see when AI becomes part of the design loop.

In the end, it's not about more tools — it's about turning momentum into mastery.

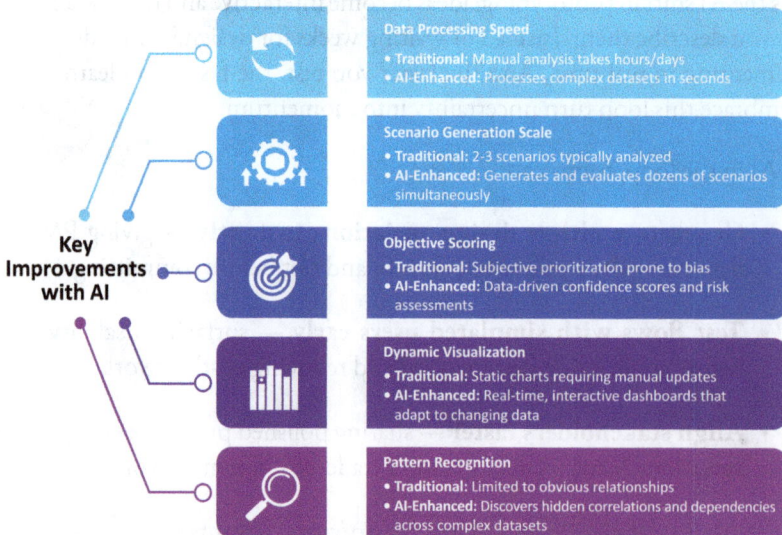

Key Improvements with AI

Data Processing Speed
- **Traditional:** Manual analysis takes hours/days
- **AI-Enhanced:** Processes complex datasets in seconds

Scenario Generation Scale
- **Traditional:** 2-3 scenarios typically analyzed
- **AI-Enhanced:** Generates and evaluates dozens of scenarios simultaneously

Objective Scoring
- **Traditional:** Subjective prioritization prone to bias
- **AI-Enhanced:** Data-driven confidence scores and risk assessments

Dynamic Visualization
- **Traditional:** Static charts requiring manual updates
- **AI-Enhanced:** Real-time, interactive dashboards that adapt to changing data

Pattern Recognition
- **Traditional:** Limited to obvious relationships
- **AI-Enhanced:** Discovers hidden correlations and dependencies across complex datasets

Collaboration, Not Replacement

AI isn't here to take over design. Its role is to clear the path so designers can focus on higher-value decisions and PMs can contribute meaningfully earlier. The goal isn't to finalize work in AI, it's to create a shared foundation for exploration so that by the time humans weigh in, the conversation is about nuance, creativity, and strategy — not catching up on first drafts. AI doesn't finish the work, it makes sure the real work starts stronger.

⚠️ Landmines and Blind Spots

AI can accelerate design, but only if you stay alert to the traps that come with it. Speed without discipline creates noise instead of lucidity, and that's where teams slip. The real edge comes from knowing when to trust the output and when to challenge it. Like any tool, AI in design comes with potential pitfalls:

- Generic outputs that lack brand personality.

- Accessibility issues that slip through without guardrails.

- Over-automation that skips crucial human critique.

AI should be embedded within a human-centered process, where oversight and refinement remain non-negotiable. Handled well, AI becomes a force multiplier, not a liability.

> **" AI doesn't replace creativity—it accelerates it by removing the friction between imagination and execution "**

From Speed to Collaboration

In a fast-moving market, learning quickly is the ultimate advantage. Turning validated insights into prototypes within hours unlocks sharper collaboration and faster progress.

- Moving from insight to prototype in hours allows teams to:

- Explore more ideas — generate and compare multiple approaches quickly.

- Test more hypotheses — validate assumptions faster and with less risk.

- Iterate with confidence — refine concepts rapidly using real feedback before committing resources.

But speed alone is not enough. The breakthrough comes when speed is paired with collaboration and inclusivity. AI helps PMs, designers, and engineers share a common view of the problem and options. Instead of debating assumptions, teams test them together and build confidence through evidence.

The future of design is about creating a collaborative, inclusive, and intelligent process that helps teams learn faster, decide smarter, and deliver with transparency. The teams who master this will set the pace for everyone else.

Prompt Library: Design & Prototyping

Wireframe Generation — AI generates wireframes in seconds so teams react to something tangible right away. Designers focus on refining instead of drafting.

Wireframe Generation

Use these prompts to quickly generate layout ideas and interface structures.

- *Design a mobile onboarding flow for a fitness app that includes goal selection, personalization, and a progress dashboard.*
- *Generate a wireframe for a dashboard that displays user analytics, recent activity, and quick actions for a SaaS admin panel.*
- *Create a homepage layout for an e-commerce site selling eco-friendly products, including hero section, featured products, and testimonials.*

UX Copywriting — AI generates microcopy, tooltips, and CTAs in context, letting teams test tone and clarity early. Copy shifts from bottleneck to accelerator.

UX Copywriting

Use these prompts to create microcopy, tooltips, and CTAs that enhance user experience.

- *Write friendly, concise microcopy for a password reset flow, including error messages and confirmation screens.*
- *Generate onboarding tooltips for a project management app that guide users through creating their first task.*
- *Create call-to-action button text variations for a pricing page targeting small business owners.*

Flow Mapping — AI maps end-to-end flows, flags dead ends, and surfaces friction before release. Teams refine paths continuously.

Flow Mapping

Use these prompts to visualize user journeys and interaction paths.

- *Create a user journey for a new user signing up for a freemium SaaS tool and upgrading to a paid plan.*
- *Map out the steps for a user booking a hotel room through a travel app, including search, filter, and checkout.*
- *Design a flow for a user reporting a bug in a mobile app, from issue discovery to confirmation.*

Design Critique — AI stress-tests accessibility, hierarchy, and usability from the start, raising baseline quality. Reviews can focus on strategy, not cleanup.

Design Critique

Use these prompts to evaluate and improve design quality and usability.

- *Review this UI layout and suggest improvements for accessibility and clarity.*
- *Evaluate the visual hierarchy of this landing page and recommend changes to improve conversion.*
- *Analyze this mobile app screen and identify any usability issues or inconsistencies with design guidelines*

Prototyping & Testing — AI spins up interactive, testable versions in minutes. Prototyping shifts from defending one idea to learning which ideas work.

> ⚙️
>
> **Prototyping & Testing**
> Use these prompts to simulate user feedback and optimize design decisions.

- *Suggest 3 variations of this checkout flow to test for conversion optimization.*
- *Generate usability test questions for a new feature that allows users to schedule recurring tasks.*
- *Simulate user feedback for a prototype of a social media sharing feature in a news app.*

Teams can spin up variations, test quickly, and adapt without delay. AI sharpens each stage of design, boosting creativity while making work faster and easier to adjust. The payoff: more options explored, blind spots caught early, and focus locked on what matters most. With AI as a partner, every iteration compounds into traction and confidence.

5 Tools to Try This Week

Figma	For wireframe generation and design suggestions
uizard	For turning prompts into UI mockups
ChatGPT	For UX copywriting
maze	For prototype testing
Notion	For design documentation

3 Prompts to Test Today

1. Design a mobile onboarding flow for a fitness app.
2. Write microcopy for a password reset flow.
3. Suggest 3 improvements to this UI layout for clarity and accessibility.

Mindset Shift to Adopt

AI removes friction between creativity and execution, but the real value is more than speed. Faster iteration means more chances to explore and refine before you commit. Treat AI as an accelerator, not autopilot. Shift from "deliverables first" to "learning first," and you raise both creative quality and strategic relevance. The PMs who win pair rapid execution with intent, using AI to amplify judgment, not replace it.

Reflective Questions

- Where in your design process do you consistently lose momentum?

- If you could test three more variations every sprint, what would you learn faster?

- How could AI flows or copy free designers to focus on creativity over first drafts?

- What's one design experiment you could run this week to prove AI speeds up iteration without sacrificing quality?

Exercise

Pick one feature from your backlog. Prototype it the way you normally would. Then create a second version using AI for wireframes, flows, and UX copy. Bring both to your next team sync and compare them side by side. Notice how the team's discussion changes when AI accelerates the first draft.

Key Takeaways

- AI removes bottlenecks, making prototyping a continuous flow.

- The edge isn't speed alone, it's exploring more directions without added cost.

- Early validation prevents wasted cycles and budget.

- AI frees designers to focus on higher-level craft and vision.

Chapter 8: Faster Launches, Smarter Execution

How AI Eliminates the Chaos

Chapter 6 sharpened the feedback loop. Now we shift to execution, turning clarity into delivery where ideas launch or stall.

Once prototypes harden into plans, the next move is shippable specs. I use AI to draft requirements fast. The full toolkit is at mach10pm.com/resources.

I worked with a PM who dreaded release week. The backlog was messy, QA lagged, every sprint a fire drill. Then the team used AI for backlog cleanup, sprint forecasting, and test automation. Delivery became predictable. That PM walked into meetings confident, backed by data and clear priorities.

Scaling delivery is not about doing more but about focus. As products grow, demands multiply. Old playbooks added headcount and process, draining momentum. AI flips that. Instead of piling on people, you build systems that absorb complexity and surface clarity.

With predictable delivery, teams stop scrambling and start shipping with conviction. Stakeholders see progress they trust, and customers feel it in every release.

Where Delivery Gets Stuck

Strong teams still hit friction moving from backlog to release. The problem is not talent, it is bottlenecks:

- QA behind

- Priorities shifting

- Release prep draining time

- Status reporting eating hours

AI handles the load so issues surface early and get fixed before slowing progress.

At a mid-stage SaaS company, release week was chaos. QA lagged, the backlog was messy, and estimates failed. After adding AI to forecast velocity and generate test cases, launches became predictable. Leadership stopped dreading release meetings and started celebrating them.

From Insight to Impact

Signals only matter if they make it into production. The advantage comes when clarity flows smoothly through sprints and releases. AI helps by:

- Cleaning up backlogs and removing duplicates

- Forecasting sprint outcomes based on team velocity

- Generating test cases and edge scenarios in minutes

- Drafting release notes and documentation automatically

With the grunt work handled, PMs and engineers can focus on judgment, strategy, and faster delivery.

AI as Your Delivery Engine

Once insights are flowing, AI pushes them through development without bogging teams down. Old execution meant manual backlog grooming, guesswork in sprint planning, and hours lost in test cases. AI removes that friction.

- **Backlog grooming:** review tickets, remove duplicates, group into themes

- **Sprint forecasting:** predict completion based on velocity history

- **Scope detection:** flag creep or bottlenecks before they stall a sprint

- **Test generation:** create edge cases in minutes for broader QA

Delivery is no longer a black box. Progress is visible, blockers surface sooner, and teams align with confidence.

How AI Enhances Development and Delivery

TASK	TRADITIONAL	AI-AUGMENTED
Backlog Grooming	Manual Review and Tagging	AI-Suggested Prioritization and Deduplication
Sprint Planning	Team-Led Estimation	AI-Assisted Capacity Planning and Velocity Forecasting
Code Review	Peer-Based	AI-Generated Suggestions and Bug Detection
QA & Testing	Manual Test Case Writing	AI-Generated Test Cases and Regression Detection
Release Notes	Written by PMs	AI-Generated Summaries from Commits and Tickets

AI compresses launch cycles from months to days

Prompting for Delivery Support

AI becomes a real ally when you direct it with precision.

Examples:

- **Backlog Grooming**
  Prompt: "Review these 50 Jira tickets and suggest duplicates or outdated items."
  Prompt: "Group these backlog items into themes or epics."

- **Sprint Planning**
  Prompt: "Given a velocity of 30 points and these 10 tasks, suggest a feasible sprint plan."
  Prompt: "Estimate effort for these user stories based on history."

- **Test Generation**
  Prompt: "Write test cases for a login flow with email validation, password reset, and 2FA."
  Prompt: "Generate edge scenarios for a checkout process with promo codes and multiple addresses."

- **Release Notes**
  Prompt: "Draft user-facing release notes from these commit messages."
  Prompt: "Summarize improvements for the customer success team."

Each one saves hours and keeps delivery flowing.

AI-Augmented Delivery Pipeline

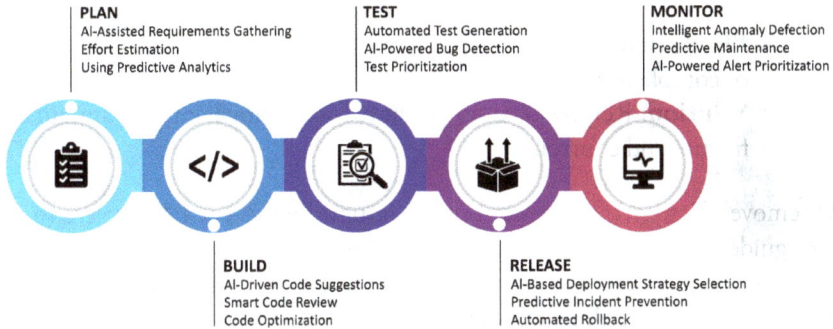

PLAN
AI-Assisted Requirements Gathering
Effort Estimation
Using Predictive Analytics

TEST
Automated Test Generation
AI-Powered Bug Detection
Test Prioritization

MONITOR
Intelligent Anomaly Defection
Predictive Maintenance
AI-Powered Alert Prioritization

BUILD
AI-Driven Code Suggestions
Smart Code Review
Code Optimization

RELEASE
AI-Based Deployment Strategy Selection
Predictive Incident Prevention
Automated Rollback

Execution Without the Chaos

AI should support delivery, not dictate it. Its role is to reduce friction, not replace judgment. Done right, it lightens the load, sharpens communication, and keeps sprints aligned. Delivery becomes more collaborative, predictable, and adaptable. The difference is immediate—less firefighting, fewer missed handoffs, and more time spent on the work that actually moves the product forward.

But speed without intention can backfire. Acceleration without alignment just creates prettier chaos. That's why the next step is understanding the guardrails that keep AI-powered delivery safe, effective, and worthy of your team's trust. This isn't about chasing speed for its own sake. It's about building a system that stays fast **and** stays sane as complexity scales.

⚠️ Landmines and Blind Spots

AI can supercharge delivery, but without discipline it can create new risks. The key is to stay intentional and pair speed with oversight.

AI strengthens delivery when applied with care:

- **Over-automation** — treating AI as autopilot instead of a tool that needs direction.
  Solution: Keep humans in the loop to ensure quality, relevance, and accountability.

- **Shallow adoption** — experimenting without embedding AI into real workflows.
  Solution: Start small, refine your processes, and scale only what actually delivers value.

- **Blind trust** — accepting outputs without validating feasibility, security, or compliance.
  Solution: Review, test, and audit before deployment to maintain confidence and control.

AI removes friction, but it won't remove responsibility. The PM's role is to validate, guide, and contextualize. You are the pilot, not the passenger. AI accelerates execution, but you remain accountable for the outcomes your team delivers—and the culture you build along the way.

Delivery as a Differentiator

In the AI era, delivery is more than execution. It is a differentiator. The teams that ship reliably, adapt quickly, and learn continuously are the ones that win. AI gives PMs leverage to:

- Align delivery with strategy

- Reduce waste and rework

- Build trust through predictable performance

The best PMs do not just ship features. They ship outcomes. They create engines that move with purpose, not panic—where progress is visible, priorities stay clear, and velocity compounds with every cycle. The advantage is confidence in execution: you are not guessing how fast or how well the team can ship. You know. With AI, velocity is no longer fragile. It becomes a capability you can trust and one competitors will struggle to match, no matter how many people they throw at the problem.

The ultimate measure of delivery is not speed but resilience. A team that can adapt to change, recover from setbacks, and still ship on time builds an edge that is hard to replicate. AI strengthens this resilience by making every cycle a learning cycle, so performance improves with each release instead of degrading under pressure. Over time, that resilience compounds into reputation. Teams that deliver consistently are seen as credible, customers learn to trust the cadence, and leaders begin to view delivery itself as a strategic weapon, not just an operational necessity.

> ## "AI doesn't just speed up delivery—it improves the quality and predictability of what gets delivered"

The Delivery Acceleration Stack

Examples of tools that give you leverage in delivery:

- Backlog hygiene audits: surface duplicates and priorities instantly.

- Sprint forecasting: predict delivery with confidence.

- Automated test generation: expand coverage without slowing down.

- Release note drafting: turn commit logs into clear updates for stake-holders and customers.

Integrated into your workflow, these shift teams from reacting to preventing problems.

Prompt Library: Development & Delivery

Execution is where strategy turns into impact. AI accelerates delivery by stripping out friction, tightening workflows, and keeping teams aligned. From backlog grooming to predictive monitoring, the right prompts turn every stage of development into a faster, smarter, and more confident process.

Whether you are cleaning up a cluttered backlog, forecasting sprints, automating QA, summarizing pull requests, managing releases, or monitoring delivery risk, these prompts show you how to pair AI's speed with your judgment for maximum impact.

Backlog Grooming

A cluttered backlog drags on decision-making. Grooming restores focus by keeping only the most relevant items in view. AI accelerates this by spotting duplicates, grouping themes, and refining user stories.

Backlog Grooming

Use these prompts to clean, prioritize, and organize your backlog.

- *Review these 50 Jira tickets and suggest which ones are duplicates, outdated, or high priority.*
- *Group these backlog items into themes or epics based on their descriptions.*
- *Identify any user stories that are missing acceptance criteria or are too vague to estimate."*

A sharp backlog keeps the team moving with clarity and confidence — no noise, just progress on what matters.

Sprint Planning & Forecasting

Planning is part art, part science. Sprint forecasting helps teams balance ambition with realism so capacity aligns with commitments. AI makes it faster by modeling scenarios and effort with data, not just gut feel.

Sprint Planning & Forecasting

Use these to support planning and capacity alignment.

- *Given our team's velocity of 30 points and these 10 tasks, suggest a feasible sprint plan.*
- *Estimate the effort level for each of these user stories based on historical data.*
- *What are the risks of including these 3 features in the upcoming sprint?*

Forecasting turns planning debates into alignment — teams commit with clarity and surface risks early.

QA & Test Automation

Quality is too often treated as a bottleneck. AI-generated test cases flip the script, broadening coverage without slowing teams down. Instead of QA lagging behind, teams can catch issues earlier, validate edge cases they would have missed, and ship with more confidence. The real win isn't just faster testing, it's creating a safety net that scales as the product grows.

QA & Test Automation

Use these to generate test cases and improve quality assurance.

- *Write test cases for a login flow that includes email validation, password reset, and 2FA.*
- *Generate edge case scenarios for a checkout process with promo codes and multiple shipping addresses.*
- *Suggest automated regression tests for a new commenting feature in a blog platform.*

Automation builds confidence — teams release faster knowing quality is baked in, not inspected at the end.

Code & PR Summarization

Pull requests connect strategy to execution, but without translation, critical updates get lost. AI bridges the gap by summarizing code and PRs in plain language. It turns dense commits into clear narratives that everyone can understand. Instead of forcing teams to decode technical details, you give them context they can trust and decisions they can make faster.

Code & PR Summarization

Use these to improve communication and understanding across teams.

- *Summarize this pull request in plain language for a non-technical stakeholder.*
- *Explain the purpose and impact of these recent code commits.*
- *Highlight any potential risks or breaking changes in this PR.*

Summaries reduce friction — engineers stay aligned, stakeholders stay informed, and transparency becomes the default.

Release Management

Releases aren't just technical milestones. They are customer moments. AI streamlines documentation and changelogs so launches feel less like chores and more like traction. It keeps teams focused on impact instead of paperwork and helps customers see progress in a way that builds trust and excitement. When every release tells a clear story, it strengthens the relationship between your product and your users.

Release Management
Use these to streamline release documentation and communication.

- *Generate user-facing release notes from these commit messages and Jira ticket summaries.*
- *Create a changelog entry for version 2.1.0 based on these updates.*
- *Summarize the key improvements and bug fixes in this release for the customer success team.*

Strong release communication builds confidence across teams and turns updates into growth opportunities.

Predictive Delivery & Monitoring

Delivery speed only matters with predictability. Predictive monitoring helps PMs spot risks early, adjust quickly, and forecast outcomes that build stakeholder trust.

Predictive Delivery & Monitoring
Use these to anticipate issues and optimize delivery flow.

- *Based on the last 3 sprints, forecast the likelihood of completing this sprint on time.*
- *Identify potential bottlenecks in our current delivery pipeline.*
- *What indicators suggest that our team's velocity is declining?*

AI handles documentation so teams stay focused on impact and customers see progress they can trust. The best PMs ship with confidence.

5 Tools to Try This Week	
⊕ **GitHub Copilot**	For code generation and refactoring
🌑 **Linear**	For sprint planning and backlog grooming
testim	For automated QA
✳ Claude	For release note generation
🔷 Jira	For delivery forecasting

3 Prompts to Test Today
1. Review these 50 Jira tickets and flag duplicates or outdated ones.
2. Generate test cases for a login flow with 2FA.
3. Summarize this pull request for a non-technical stakeholder.

Mindset Shift to Adopt

AI reduces friction between planning and execution. The opportunity isn't just the time saved—it's how you reinvest it. Use the gains to expand QA coverage, strengthen alignment, and improve quality. Think of AI as a delivery accelerator, not a replacement.

Reflective Questions

- Which part of your delivery pipeline slows you down most: backlog, planning, QA, or release?

- How could AI automation free engineers and QA to focus on higher-value work?

- Where could faster feedback loops improve confidence and reduce wasted effort?

- What risks could increase if you speed up delivery without improving quality?

Exercise

To see the impact clearly, do not just theorize. Run AI side by side with your current approach to reveal where it adds value and how it changes the way you work.

- **Baseline:** Follow your current process without AI.

- **AI-Augmented:** Use AI to suggest backlog priorities and generate test cases.

- **Identify:** Document one or two AI-driven practices you can integrate right away.

Key Takeaways

- AI compresses delivery cycles by automating steps and reducing delays.

- Structured prompts improve grooming, planning, QA, and docs.

- Predictive monitoring keeps delivery proactive by surfacing risks early.

Chapter 9: Dashboards Don't Win Markets

How AI Turns Insight into Acceleration

I worked with a PM who hit every KPI on the dashboard: traffic, signups, engagement. On paper, it looked like success. But three months later churn spiked and revenue flatlined. The team wasn't steering the business; they were staring at the gauges, celebrating green lights while the engine was already sputtering.

That's the danger when metrics become the finish line instead of the starting point. Dashboards are gauges. Engines win races. They show speed, not direction. Too many PMs mistake activity for progress.

The advantage comes when you treat metrics as navigation. Data should move you forward, not decorate a slide. AI makes that possible. It turns metrics from static snapshots into living systems that surface shifts, highlight risks, model scenarios, and guide decisions in real-time. Instead of asking "What happened?" you start asking "What is about to happen, and how should we respond?"

Momentum comes from connecting signal to action. AI tightens that loop so you learn faster, respond sooner, and move ahead of competitors still stuck in dashboard debates.

Why Dashboards Lie

Even strong PMs hit walls with data:

- **Data overload** – endless dashboards but little clarity.

- **Delayed insights** – reports land after the moment has passed.

- **Siloed metrics** – scattered across tools and teams with no unified view.

- **Reactive decisions** – trends spotted only after they've peaked.

The result: slow growth, diluted focus, and missed chances.

AI Turns Analytics Into Advantage

AI shifts analytics from reactive to proactive. Instead of staring into the rearview mirror, you can see the road ahead, anticipate curves, and adjust before you get there. This isn't just faster, it moves you from reporting on the past to actively shaping the future.

AI can:

- Surface trends the moment they emerge, giving you first-mover advantage.

- Correlate metrics to uncover hidden growth levers.

- Model the impact of pricing tweaks, feature launches, or campaigns before they happen.

- Spot risks early, from churn signals to operational bottlenecks, so you act before they escalate.

With AI in place, you're no longer chasing the game. You're setting the pace, shaping the field, and forcing others to catch up.

AI-Augmented Growth Stack

INPUTS (DATA)	AI PROCESSING	OUTPUTS (INSIGHTS, ACTIONS, FORECASTS)
Customer Data	Data Ingestion & Cleaning	Growth Opportunities
Product Usage	Feature Engineering	Automated Campaigns
Market Signals	Predictive Modeling	Churn Predictions
Internal Metrics	Natural Language Processing	Personalization Strategies
Third-Party Data	Recommendation Systems	Revenue Forecasts

AI shifts analytics from reporting the past to shaping the future

Gears of the Growth Engine

Your AI analytics edge comes from pairing the right technology with disciplined use. These tools are the gears in your growth engine. Choose them carefully, integrate them deeply, and make their use second nature across the team.

Tools for Metrics & Growth

Tool	Use Case
mixpanel + ˙AI	Automated insights, anomaly detection, funnel analysis
Amplitude + ˙AI	User behavior analytics, cohort discovery, personalized recommendations
Google Analytics + Gemini	Predictive metrics, trend forecasting
pendo + ˙AI	In-app product usage insights, churn risk prediction
Optimizely + ˙AI	Optimizely + AI Automated A/B and multivariate testing for growth experiments
ChartMogul + ˙AI	SaaS metrics automation (MRR, churn, LTV) with AI-powered forecasts

Dashboards track. AI predicts

Questions That Move Markets

Prompts are the steering wheel of your AI analytics process. Tools give you horsepower, but prompts determine where you go. The right questions open doors. Vague ones just add noise, leaving you stuck in dashboard fog. Strong prompts can reveal hidden correlations, highlight churn risks, or surface untapped segments. Weak prompts leave you staring at dashboards with no direction.

Sharpening this skill matters. Be precise, outcome-driven, and always tie prompts to business goals. For example:

- **Bad Prompt:** *Tell me about customer data.* → Too vague, mostly noise.

- **Good Prompt:** *Identify the top three drivers of churn among mid-market customers over the past two quarters, and suggest where to intervene first.* → Specific, actionable, and tied to growth.

Avoid vague or overloaded prompts, and never treat AI outputs as unquestionable truth. With the right discipline, prompting turns analytics into a growth engine that sharpens decisions, drives alignment, and keeps teams focused on what actually matters.

Prompting for Analytics & Growth

Here are examples of prompts PMs can use to extract insights and drive action:

Funnel Optimization

Analyze this signup funnel and identify the biggest drop-off points and possible causes.

Retention Analysis

Compare retention rates across three user cohorts and suggest actions to improve the lowest-performing group.

Experiment Design

Propose three A/B test ideas to improve conversion on our pricing page, including success metrics.

Forecasting

Based on the last 6 months of user growth, forecast our active user count for the next quarter.

Executive Reporting

Summarize our Q2 product performance in under 300 words, highlighting key wins, losses, and next steps.

From Metrics to Strategy

AI helps PMs move beyond tracking KPIs to strategic decision-making:

- Prioritize growth levers based on predictive impact
- Personalize onboarding or messaging based on user behavior
- Simulate outcomes of pricing or feature changes

This turns analytics from a rearview mirror into a navigation system.

Make the Numbers Move the Needle

The payoff comes when analytics inform strategy, not just measurement. Too often, teams track KPIs without tying them to decisions. AI closes that gap, turning data into predictive guidance.

With AI, you can:

- Prioritize growth levers based on predicted impact, not instinct.

- Personalize onboarding and messaging around emerging behavioral patterns.

- Run "what if" scenarios before committing resources.

When metrics inform strategy, every decision is intentional and evidence-backed.

Strong prompts light the spark — this is the engine that turns your data into growth at MACH-10 speed.

⚠️ Landmines and Blind Spots

AI analytics is powerful, but not infallible. The danger is treating outputs as truth. Flawed or biased data leads to flawed recommendations, and AI will happily deliver convincing—but meaningless—connections.

Treat AI as a trusted advisor, not an authority. Pressure-test conclusions with domain expertise and judgment before acting, especially on high-stakes moves. Watch for:

- **Correlation vs. causation** – Movement together doesn't always mean cause.

- **Data quality issues** – Garbage in, garbage out still applies.

- **Overfitting** – Models trained too narrowly can fail when conditions shift.

Great PMs use AI to generate hypotheses, not verdicts. You remain the interpreter, asking, "Does this make sense for our product, market, and users?"

Compounding Growth Loops

Too many teams chase "growth hacks" that spike metrics and fade. AI lets you build a growth engine that compounds instead of burning out. The shift is from tricks to systems that adapt, learn, and get stronger over time.

The system includes:

- **Continuous learning:** Growth adjusts in real-time to user behavior and market signals.

- **Rapid experimentation:** Short feedback loops make testing easy and push bolder ideas.

- **Cross-functional alignment:** Product, marketing, and ops move on the same evidence, not competing stories.

- **Automated optimization:** AI tunes campaigns, pricing, and messaging at scale, compounding results without extra overhead.

When these pieces connect, growth stops being one-off campaigns and becomes a living system. Each test sharpens the next, each win compounds advantage, and over time the system itself becomes the moat while competitors keep chasing hacks.

Signals from the Edge

A mid-market Fintech firm ditched one-off campaigns for an AI-driven system that ran nonstop experiments across onboarding, pricing, and messaging. AI flagged funnel friction, suggested copy tweaks, and modeled churn in real time. In two quarters they tripled experiment volume without extra headcount, turning short spikes into durable growth.

The lesson: lasting growth comes from systems that learn and adapt, not one-off wins.

> **" AI doesn't just show you what happened—it helps you understand why, and what to do next "**

Prompt Library: Metrics, Analytics & Growth

AI turns raw data into actionable insight, giving PMs the clarity to drive growth with speed, precision, and confidence. It separates the noise from the signal so you can focus on what actually moves the business forward. The result is faster decisions, sharper execution, and growth that compounds.

Funnel & Conversion Analysis

Growth depends on a frictionless user journey. Funnels show where it breaks, and AI pinpoints bottlenecks so you can design smoother paths with precision.

Funnel & Conversion Analysis

Use these prompts to identify drop-offs and optimize user journeys.

- *Analyze this signup funnel and identify the biggest drop-off points and possible causes.*
- *Compare conversion rates across mobile and desktop users for our onboarding flow.*
- *Suggest 3 improvements to increase conversion from free trial to paid plan.*

Improving funnels isn't about nudging numbers. It's about respecting the user's time and intent. Each optimized step builds trust and momentum, turning curiosity into commitment. With AI driving your conversion strategy, you're not just optimizing a flow. You're designing a journey worth completing—one that feels seamless, personal, and aligned with what users came to accomplish. Done right, the result isn't just higher conversion. It's loyalty that compounds.

Experimentation & A/B Testing

If data is the compass, experimentation is the map. A/B testing takes the guesswork out of growth, allowing you to validate assumptions with evidence instead of opinions. AI doesn't just speed up test design and analysis—it makes experimentation scalable, so learning compounds faster than your competitors.

Experimentation & A/B Testing

Use these to design, analyze, and iterate on growth experiments.

- *Propose three A/B test ideas to improve conversion on our pricing page, including success metrics.*
- *Analyze the results of this A/B test and determine if the difference is statistically significant.*
- *Suggest follow-up experiments based on this failed test result.*

Every test teaches you something, even when it fails. The point isn't to be right every time, but to learn continuously and iterate boldly. By pairing A/B testing with AI, you shift experimentation from an occasional tactic into an organizational habit that drives relentless improvement.

Reporting & Stakeholder Communication

Strong insights only drive impact when they are communicated clearly. The best PMs are storytellers who turn data into narratives that inspire confidence and drive alignment. Reporting is not busywork; it is a leadership skill. Done well, it builds credibility and ensures teams move in the same direction.

Reporting & Stakeholder Communication

Use these to summarize insights and align teams.

- *Summarize our Q2 product performance in under 300 words, highlighting key wins, losses, and next steps.*
- *Create a slide-ready summary of our top 5 growth metrics and what they mean.*
- *Write a weekly product analytics update for the leadership team.*

Reporting is the bridge between insight and influence. Crisp, actionable updates turn raw data into stories that drive decisions. When you deliver them consistently, you stop being just the person with numbers and start becoming the leader shaping direction. AI can help speed up the craft—summarizing research, highlighting anomalies, surfacing trends—but the real leverage comes from how those updates build momentum and trust. Each report becomes a signal of clarity, alignment, and intent, and over time the habit compounds into credibility that carries across the organization.

Forecasting & Predictive Insights

Analytics often looks backward. Forecasting flips that. Instead of reacting, you anticipate. AI sharpens foresight, projects growth, tests scenarios, and lowers the risk of blind bets.

Forecasting & Predictive Insights
Use these to anticipate trends and plan proactively.

- *Based on the last 6 months of user growth, forecast our active user count for the next quarter.*
- *Predict the impact of a 10% increase in onboarding completion on monthly revenue.*
- *What are the leading indicators of a successful product launch based on past data?*

Predictive insights shift how you lead. You stop reacting and start shaping outcomes before they unfold. The PMs who master this aren't reporting the game, they are calling the next play. Dashboards track the past. Systems create the future.

5 Tools to Try This Week

mixpanel	For funnel and retention analysis
Amplitude	For behavioral segmentation
GrowthBook	For experimentation and A/B testing
ChatGPT	For executive summaries
Notion	For growth dashboards

3 Prompts to Test Today

1. Analyze this signup funnel and identify drop-off points.
2. Compare retention across three user cohorts.
3. Forecast active users for next quarter based on current trends.

Mindset Shift to Adopt

AI reframes analytics: not just "What happened?" but "Why did it happen?" and "What's next?" It transforms analytics from a rearview mirror into a steering wheel for strategy, letting leaders anticipate turns instead of reacting late.

Analytics stops being a monthly chore and becomes a daily driver of decisions. It connects discovery, design, and delivery into one loop, with AI's speed fueling higher-value work:

- Refining hypotheses with real data instead of gut feel

- Stress-testing growth levers before costly bets

- Shaping roadmaps with signals instead of surprises

The payoff is confidence and control. Leaders gain clarity faster, teams align sooner, and organizations move with greater purpose. Decisions carry more weight because they're anchored in fresh evidence, not stale reports. AI helps you stop reporting the news and start shaping it, turning analytics into an engine for trust, momentum, and advantage.

Reflective Questions

- Which metrics in your product reports arrive too late to be useful?

- Where are you still drowning in dashboards instead of acting on signals?

- How could predictive insights change your next big bet?

Exercise

Select a recent product metric or growth experiment and run the data through an AI analytics tool such as Mixpanel AI or Amplitude AI. Compare the AI's findings with your team's conclusions:

1. **Align** – Note where the AI surfaced the same insights your team did.

2. **Diverge** – Identify where the AI spotted something your team missed (or vice versa).

3. **Decide** – Document which actions each approach suggests, then determine which path offers the highest potential upside.

Key Takeaways

Most PMs drown in dashboards, chasing metrics too late. MACH-10 PMs flip the script, using AI as a forward-looking engine that turns analytics into strategy. They don't just report the news—they make it by anticipating what's next and acting early. AI is how you stop analyzing the past and start shaping the future.

AI enables PMs to:
- Anticipate customer needs before they surface

- Catch risks before they become churn

- Test big bets before burning capital

- Build roadmaps that adapt in real-time

With this shift: you're not a passenger reading the dashboard. You're in the driver's seat, steering at MACH-10.

Chapter 10: Stop Collecting Tools. Start Orchestrating Them

Your Stack Isn't a Junk Drawer

In Chapter 9, we saw how AI turns metrics into real-time navigation. Now it is about what to equip yourself with. As a MACH-10 PM, tools are not conveniences. They are the infrastructure that drives speed and adaptability.

I worked with a PM who had 15 AI tools bookmarked. On paper it looked impressive. In practice, nothing connected and time was lost switching tabs. Another team ran half as many tools but wired them together. Feedback shaped prioritization, prototypes updated overnight, and analytics fueled the roadmap.

At a SaaS company I worked with, the team had seven tools for planning and tracking. Meetings became reconciliation exercises instead of decision making. When they pulled everything into one workflow and added AI, alignment finally stuck.

Your stack is not a shopping list. It is your runway. The right mix determines how fast you take off and how confidently you adjust. Success comes from curating tools that fit your workflow, amplify strengths, and cover blind spots.

AI has created an explosion of platforms. Stacking them without intention creates waste. The PMs moving fastest are not chasing every platform. They are orchestrating the ones that matter, aligning them into a system that multiplies their impact.

The MACH-10 Orchestration Framework

Stop collecting tools. Start orchestrating them with this simple flow:

1. **Map the Work** – Identify your critical workflows across discovery, design, delivery, and measurement.

2. **Match the Tools** – Assign one primary AI tool (and a backup) to each workflow stage.

3. **Integrate, Don't Isolate** – Connect your tools so insights flow across the stack, not in silos.

4. **Standardize Prompts & Processes** – Document the high-impact prompts, templates, and best practices for your team.

5. **Review & Refine** – Every quarter, audit your stack. Drop what's not adding leverage and double down on what is.

The MACH-10 Orchestration Flow

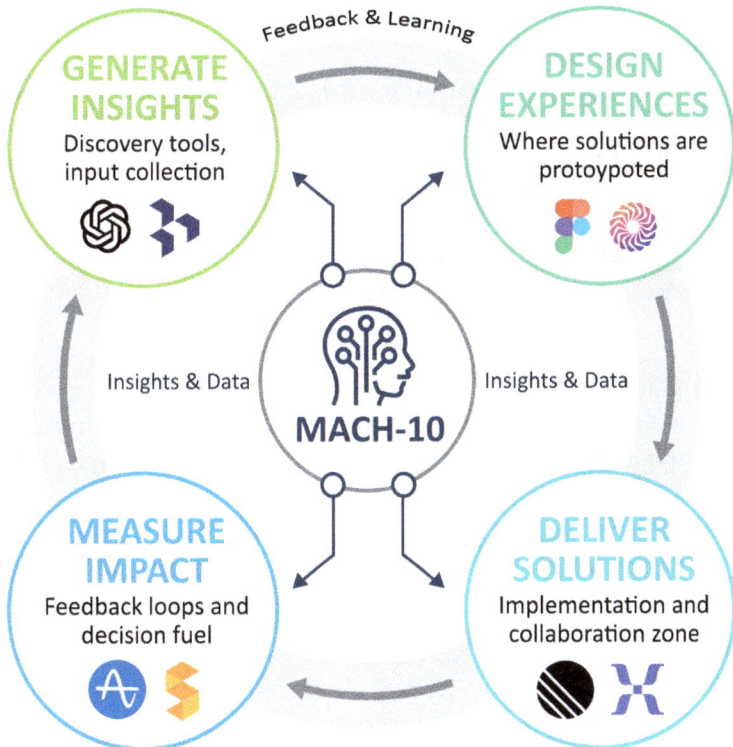

Tools don't win alone. Orchestration creates advantage

When you apply this framework, your workflow stops feeling fragmented. Discovery informs design in real time. Delivery connects seamlessly with measurement. Feedback loops tighten, and decisions get sharper not because you have more tools, but because you are using the right ones, in the right way, at the right time.

This shift creates confidence and momentum across the team. Instead of chasing disconnected signals, everyone works from the same source of truth. Each decision compounds into the next, and your stack evolves from a loose set of apps into an operating system that drives clarity, speed, and results.

AI Tools That Power the PM Workflow

The PM role spans from discovery to delivery, so your stack should reflect that range. Think of it not as a menu but as a map, showing where each tool creates leverage.

ChatGPT / Claude
Partners for ideation, synthesis, and writing. Draft a PRD, cluster feedback, or role-play as your user. They flex across nearly every PM task, making them the core of your toolkit.

Figma AI
Speeds design with auto-layouts, copy suggestions, and interaction flows. It gives PMs and designers shared ground to test options before engineering steps in.

Productboard AI
Turns raw feedback into prioritization signals. It clusters inputs, scores features, and drafts roadmaps you can defend with confidence.

GitHub Copilot
Generates and explains code. You may not ship lines yourself, but Copilot sharpens technical conversations and trade-off decisions.

Mixpanel AI
Simplifies analytics by turning complexity into plain language. Funnels, churn risks, and predictive signals surface without a data scientist.

The value isn't in one tool but in how they work together. The next graphic shows how they slot into the PM workflow, keeping you covered from spark to ship. Seeing the tools in isolation is useful, but the real leverage comes from mapping them to the functions you perform every day. Each stage of product management, from sparking ideas to measuring growth, has AI partners that lighten the load and sharpen outcomes.

When your tools are wired with intent, they create momentum that builds with every cycle. The following chart shows how these tools line up by function, so you can spot where your stack is strong, where it has gaps, and how to weave them together into a connected system.

AI Tools by PM Function

FUNCTION	TOOL LOGOS	USE CASES
Ideation & Writing	ChatGPT, Claude	Brainstormning, specs, meeting ntes
User Research	Dovetail, EnjoyHQ	Interview analysis feedback synthesis
Design & Prototyping	Figma AI, Gailleo AI	Wireframes, UX copy, flow mapping
Roadmapping & Prioritization	Productboard AI, Aha!	Feature scoring scenario planning
Development & Delivery	GitHub, Linear, Testim	Code suggestions sprint planning, QA
Analytics & Growth	Mixpanel, Amplitude AI	Funnel analysis retention modeling

A system of tools becomes a system of truth

What matters is not just knowing which tools exist, but weaving them into your daily flow. That's where integration tips come in — turning a scattered toolkit into a connected system.

Make the Tools Work for You

Owning tools is not the same as using them well. To move from occasional dabbling to daily leverage:

- **Start small.** Pick one workflow like spec writing and layer AI in until it sticks.

- **Customize prompts.** Generic inputs give generic outputs. Tune them to your product and customers.

- **Collaborate.** Share AI outputs with your team and refine them together.

- **Stay ethical.** Protect data, respect privacy, and make fairness non-negotiable.

Getting value from AI isn't about adding tools to your workflow. It's about turning them into a system you can trust. Mastery comes from depth, not variety. When you use AI to write specs, synthesize research, or forecast sprints, tools stop feeling experimental and start acting like teammates. That's when AI shifts from novelty to default.

Once you're confident at the workflow level, build a stack. Not a loose set of apps, but a system of tools that span the product lifecycle. Real leverage comes when insights drive prioritization, prototypes evolve quickly, and analytics guide decisions.

Building Your Stack

Think of your AI toolkit like your loadout before a mission. Each piece has a purpose, and together they cover every angle of the product lifecycle. A balanced stack for a mid-sized PM org might look like this:

- **Writing & Ideation (ChatGPT + Notion AI):** Draft specs, cluster feedback, and brainstorm without the blank-page drag.

- **Research & Feedback (Dovetail + Sprig):** Turn raw interviews and surveys into crisp insights that shape real decisions.

- **Design & UX (Figma AI + Uizard):** Explore variations in hours, not weeks, and give designers and PMs shared ground to iterate fast.

- **Roadmapping (Productboard AI):** Score features, model scenarios, and draft roadmaps you can defend with clarity.

- **Delivery (GitHub Copilot + Linear AI):** Automate test cases, flag risks, and sharpen conversations with engineering.

- **Analytics (Mixpanel AI + GrowthBook):** Cut through complexity with plain-English answers and predictive signals you can act on.

A solo PM's stack will be leaner. A 15-person org may wire in deeper integrations. The point isn't tool count. The point is leverage. Every time you feel bogged down, ask yourself: *could this be AI-native?*

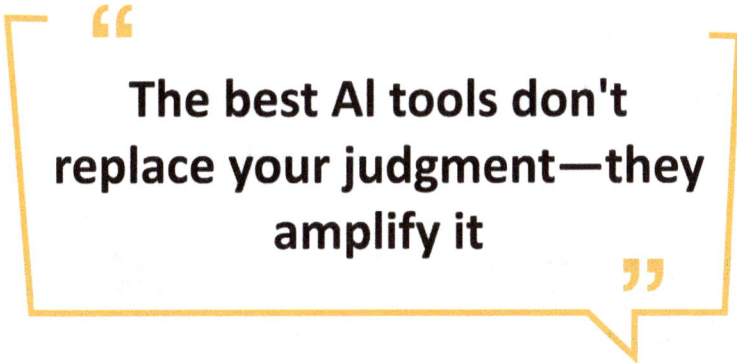

> **The best AI tools don't replace your judgment—they amplify it**

From Apps to an Operating System

The magic is not in the tools. It is in the connections.

Analytics flags a drop in engagement → prototypes update onboarding → AI drafts new copy → CRM pushes updates to marketing → a campaign launches overnight. One signal triggers coordinated action across the stack without handoffs.

That is compounding efficiency. Each connection saves time, prevents mistakes, and builds momentum. Multiply that across sprints and the gains add up to weeks of execution saved. The real payoff is not just speed but alignment, because when systems are wired together, the team sees the same signals and responds in sync.

Momentum compounds further when orchestration becomes habit. Instead of chasing scattered updates, you are operating from a living system that learns, adapts, and accelerates with every cycle. What once felt like a set of disconnected apps now works like a single operating system, and that is when the stack starts working for you instead of the other way around.

Tools do not make momentum. Orchestration does.

5 Tools to Try This Week	
ChatGPT	For ideation and writing
Figma	For design iteration
productboard	For roadmap generation
GitHub Copilot	For technical collaboration
mixpanel	For predictive analytics

3 Prompts to Test Today
1. List 5 AI tools that align with my product goals.
2. What are the top 3 use cases for AI in my current workflow?
3. Suggest an AI tool for each phase of the product lifecycle.

Mindset Shift to Adopt

Don't treat your stack like a junk drawer of apps. Treat it like an operating system. Depth over breadth. Mastery over novelty. The best MACH-10 PMs build durable workflows, refine the seams between them, and make those systems repeatable across teams.

Your goal is not just to have tools. It is to orchestrate them into a system where insight turns into action in minutes, not weeks. That's not automation for its own sake. It's leverage that compounds, accelerates alignment, and frees you to focus where judgment matters most.

Reflective Questions

- How well are your current AI tools integrated into your daily product workflow?

- Which part of your PM lifecycle has the weakest AI support right now?

- If you had to cut your stack in half, which tools would you keep and why?

- How do you decide when a new tool is worth adding?

Exercise

Audit your current stack using the following method:

- **Green:** Tools you already use effectively and consistently

- **Yellow:** Tools you use but haven't fully integrated into daily or weekly processes

- **Red:** Tools that fill a capability gap but are not yet in use

Choose one **Yellow** tool and commit to fully integrating it into your process for the next 30 days. Select one **Red** tool, run a structured trial, and track changes in speed, clarity, and quality.

Key Takeaways

- AI tools deliver the most value when built into repeatable workflows.

- Depth beats breadth. A smaller, connected stack outperforms scattered familiarity.

- The right stack turns AI from assistant to infrastructure.

- As your product scales, your stack should scale with it. Review and refine regularly.

Chapter 11: The Prompt is the Product

How to Write Prompts That Deliver Results

In Chapter 10, you built your AI stack—the tools that fuel speed, precision, and adaptability. But tools alone don't create value. Performance comes from how you direct them.

Prompt engineering is the bridge between intention and output. It turns AI into a partner that understands goals, context, and constraints. Weak prompts waste cycles. Strong ones land fast—clear, useful, cleaner than expected.

At a SaaS startup, a PM asked an AI model to rank features. It pushed a flashy integration to the top. When she pressed for reasons, the AI admitted it had weighted social mentions too heavily. That flipped the conversation: sentiment didn't equal revenue. The team shifted course.

Prompt engineering shapes speed, quality, and creativity. Whether clustering feedback, scoring features, or drafting a brief, the way you frame the request determines how close you get to a usable draft. Treat prompting like setting the stage. If you give context and intent, the system has what it needs to perform.

When I first tested GPT-4, I asked for a "detailed product spec." It came back with 1,200 words of fluff—no structure, no context. That was on me. I'd asked for "detailed," not "usable." That's when it clicked: prompting is briefing. And briefing is everything.

Turning Prompts Into Power

By this point, you've already seen prompts at work. The question isn't *what is a prompt* — it's *how do you use them with intent and consistency.*

Strong prompts are patterns. They shape the output, set the structure, and focus the system on what matters. PMs who treat prompts as reusable building blocks create leverage that compounds across every project.

- A precise question surfaces risks before they blindside the team.

- A structured task turns messy research into something you can act on.

- A role-play scenario pressure-tests choices in ways a static framework never will.

The goal isn't to dabble with prompts. It's to master them. Vague inputs give you noise. Clear prompts give you insight you can defend. The best PMs build a library, refine it over time, and treat it like living code that evolves with the product and the market.

The PM's Prompting Edge

Prompting isn't a parlor trick. It's a skill that separates dabblers from PMs who use AI as a real advantage.

The difference shows up fast:

- Weak prompts give you vague lists that could apply to any product.

- Strong prompts surface patterns, trade-offs, and options tied to your actual goals.

- Mastered prompts create reusable workflows that scale across the team.

For PMs, this is the new literacy. If you can shape inputs with clarity, you'll get outputs you can defend in a roadmap review, a board meeting, or a sprint planning session. That's the bar. Prompts aren't just questions to a system. They're the levers that turn AI into part of your operating model.

The Anatomy of a Killer Prompt

The difference between AI that sounds like a rookie and AI that sounds like a pro starts with how you frame the ask. Before diving into practical examples, it's worth anchoring on the three elements that consistently separate vague AI answers from outputs that feel like a seasoned PM wrote them:

- **Context** – Supply relevant background so the AI understands your environment, constraints, and priorities. Without it, you'll get generic answers that could apply to any product in any industry.

- **Clarity** – State exactly what you want, in the format you want it. Ambiguity leads to wandering responses. Specificity drives precision.

- **Constraints** – Limit the scope to avoid over-generalized or irrelevant results. This could be timeframes, character limits, or a defined output structure.

These principles apply whether you're prompting for strategy, copy, analytics, or design. They're also the foundation for scaling AI across a team. A good prompt is repeatable, teachable, and easy to adapt across workflows. **Done well, they turn AI from a clever assistant into a reliable extension of your judgment.**

Prompt Clinic: Weak vs Deadly

Weak prompts waste time. Strong prompts give you insight you can act on. The best prompts also ask why, exposing the assumptions, trade-offs, and signals behind the output. That is where deeper insight lives. Here are four real examples that show the difference.

1. Discovery and Research

- **Bad Prompt:** *"Summarize customer feedback."*

  - **Why it fails:** Too broad. No timeframe, no scope, no direction. The output is a generic list anyone could guess.

- **Good Prompt:** *"Cluster the top pain points from 1,200 support tickets from the past quarter. Show frequency, include a few quotes, and map each issue to activation, retention, or upsell. Rank them by business impact."*

  - **Why it works:** Clear scope and structure. Instead of noise, you get prioritized signals.

2. Roadmap and Prioritization

- **Bad Prompt:** *"Give me some ideas for our roadmap."*

  - **Why it fails:** Aimless. You'll get a random feature list with no relevance to your strategy.

- **Good Prompt:** *"We are a mid-market SaaS company focused on improving retention among SMBs in Q3. Generate five initiatives aimed at retention. For each, include user impact, engineering effort, key risks, and signals to track post-launch. Present as a ranked table with projected ROI."*

- **Why it works:** Adds context, constraints, and format. The result is a structured plan you can defend in a meeting.

3. Design and UX

- **Bad Prompt:** *"Help me improve onboarding."*

  - **Why it fails:** No product, no users, no success criteria. You'll get generic clichés.

- **Good Prompt:** "Compare three onboarding flows for a mobile budgeting app targeting Gen Z with multiple income streams: interactive tutorial, gamified progress tracker, and AI-driven personalization. For each, estimate conversion lift, drop-off risk, and engineering effort. Summarize in a table and recommend the best option, citing competitor benchmarks. Include a brief rationale explaining why your top choice aligns with Gen Z user behavior."

  - **Why it works:** Frames concrete options, defines criteria, and asks for a recommendation.

4. Analytics and Growth

- **Bad Prompt:** *"Tell me about churn."*

  - **Why it fails:** That's not a prompt, it's a definition request. You'll get theory, not insight.

- **Good Prompt:** "Identify the top three churn drivers among mid-market customers over the last two quarters using product logs and survey data. For each, propose one retention experiment with success metrics. Summarize in a short growth plan. Highlight which experiment is likely to deliver the fastest measurable impact."

  - **Why it works:** Tight scope, defined data, and an actionable output. Instead of concepts, you get a plan. It reads like marching orders to a capable teammate, not a vague wish tossed into the void. This isn't just clear—it's actionable. A prompt like this gives the AI direction, scope, and a standard for what "good" looks like.

When prompts deliver like this, they stop being one-off hacks and start becoming part of how your team thinks.

Why This Matters

Prompts create compounding impact. Weak ones waste cycles. Strong ones save hours, refine judgment, and align teams faster. Think of it like briefing a teammate: vague requests get vague work, while clear context and outcomes drive results. AI works the same way. The clearer your ask, the greater the advantage you create.

The Prompt Playbook

Effective prompting starts with a clear goal, a focused ask, and a structure that guides the system to useful output. The categories that follow are high-value use cases every PM should know. The real skill is matching the right prompt to the right moment. Get that timing wrong and you'll drown in noise. Nail it, and you turn AI into a force multiplier that feels almost unfair.

Done well, prompting becomes less about tricks and more about creating repeatable habits that raise the quality of decisions across the team.

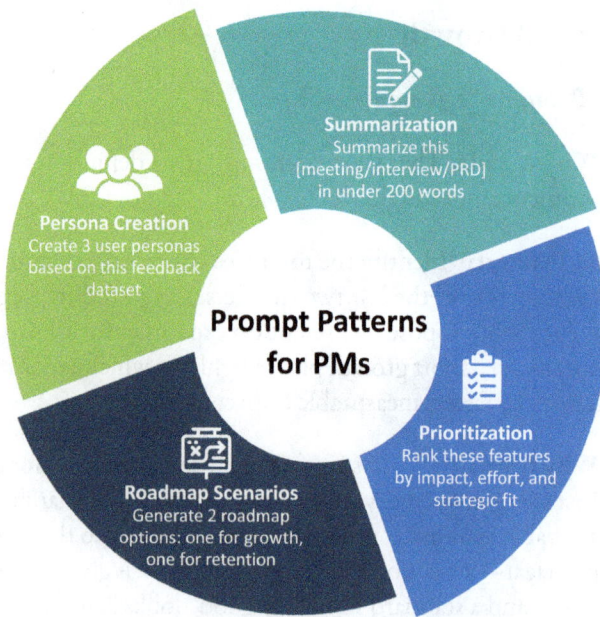

Prompt Patterns for PMs

Summarization — Summarize this [meeting/interview/PRD] in under 200 words

Persona Creation — Create 3 user personas based on this feedback dataset

Roadmap Scenarios — Generate 2 roadmap options: one for growth, one for retention

Prioritization — Rank these features by impact, effort, and strategic fit

Sharp prompts create sharper outcomes

From Patterns to Power Moves

Patterns are your starting moves, but your technique must evolve. It is time to move beyond templates into methods that help the AI think better, not just faster. Techniques like few-shot prompting and chain-of-thought reasoning unlock deeper insight, stronger outputs, and more strategic thinking. This is where prompting shifts from mechanical to intentional.

Building Prompt Instincts

Patterns get you moving, but instincts make you dangerous. The best PMs shape prompts in real time. They sense when context is thin, when an ask will miss, and how to guide the AI through the trade-offs they would make themselves.

This is not about memorizing formats. It is about developing feel. You learn to spot gaps fast, sharpen vague asks, and push until the output is crisp. Over time it becomes automatic. You stop wrestling with the tool and start thinking through it.

High-Precision Prompting

The basics get you functional. These techniques make you formidable. Advanced prompting lets you shape outputs with precision, apply constraints, and adapt them to complex product scenarios. It's the difference between getting an answer and getting the right answer.

Few-Shot Prompting

Provide examples of what you want:

Here are 2 examples of good release notes.
Now write one for this release

Chain-of-Thought Prompting

Ask the AI to reason step-by-step:

List the pros and cons of each option before making a recommendation

Role + Constraint Prompting

Add constraints to guide output:

Act as a PM at a fintech startup. Suggest 3 features under 2 weeks of dev time

Advanced Prompting Techniques

Prompts are leverage — small inputs, big impact

Precision Prompts for High-Impact AI

Prompts are the steering wheel of AI. Get them right, and you unlock leverage that feels almost unfair. Get them wrong, and you're stuck editing vague, generic answers that waste cycles instead of saving them. The following pages are a curated library of precision prompts—organized around core PM workflows—that you can plug directly into your own process. Think of them as accelerators, not scripts. You'll adapt them to your team, your goals, and your product, but the scaffolding here gets you 80% of the way to a strong first draft.

1. Discovery & Research

Discovery is where every product journey begins, but it's also where bottlenecks pile up. Reports are slow, interviews are limited, and survey data takes weeks to code. AI compresses that cycle. With the right prompts, you can scan markets, cluster feedback, and surface insights in hours instead of months.

Discovery & Research
When you're exploring a market or trying to understand customer needs, speed matters. You don't have time to sift through endless reports or manually categorize survey data. Well-structured prompts turn AI into a rapid research assistant that can process large volumes of information and surface the most relevant insights for you.

- *Summarize the top five industry trends affecting [industry] over the next 12 months. Include sources.*
- *Identify 10 customer pain points from these survey responses and group them into themes.*
- *Analyze this competitor's feature set and suggest three opportunities for differentiation.*

These prompts work because they go beyond raw information. They direct AI to synthesize, categorize, and highlight the patterns you can act on. Instead of noise, you get understanding—and instead of starting from scratch, you begin with a structured view of the landscape.

2. Roadmapping & Prioritization

Once you've gathered insights, the next challenge is focus. Which opportunities deliver the most impact in the shortest time? AI helps cut through bias and gut feel by exposing trade-offs, simulating options, and clarifying the rationale.

Roadmaps are where politics, opinions, and scarce resources collide. AI won't eliminate the hard calls, but it makes them clearer—and gives you a stronger case for the decisions you make.

Roadmapping & Prioritization
Once you've gathered insights, the next challenge is focus. Which opportunities will deliver the most impact in the shortest time? AI can help cut through bias and gut-feel decision-making, offering a structured perspective that supports strategic prioritization.

- *Given this backlog, prioritize the top 5 items for next quarter based on impact and effort. Explain your reasoning.*
- *Propose three new features for our product that align with our current strategic goals and target market.*
- *Suggest a phased rollout plan for this feature to minimize risk while maximizing early feedback.*

By including context—impact, effort, goals—these prompts force AI to move past laundry lists and into structured recommendations. The result is not a machine-made plan but a decision space that's easier to defend, adapt, and align your team around.

3. Design & UX

Design is often where good intentions slow down. Iterations take too long, UX copy gets delayed, and accessibility checks are left to the end. AI helps remove that friction. With the right prompts, you can accelerate exploration, generate testable variations, and spot issues earlier.

Design & UX
While you'll still rely on your design team for the final product, AI can help accelerate concept development, generate variations for testing, and spot issues earlier in the process. The goal here is to reduce bottlenecks, not replace expertise.

- *Generate three alternative user flows for the onboarding process to reduce drop-off.*
- *Suggest accessibility improvements for this mobile app design.*
- *Write UX copy for this error message in a friendly but professional tone.*

Of course, AI can't know your brand or customers as well as you do. But it can offer perspectives you might not have considered and drafts you can refine faster. The real advantage is iteration speed.

4. Metrics & Analysis

Data by itself is just noise. The challenge is spotting the signal before it's too late. AI turns your dashboards into a proactive analyst, surfacing patterns, risks, and opportunities you might otherwise miss.

Metrics & Analysis
Data without interpretation is just noise. The right prompts turn AI into a proactive analyst, surfacing opportunities and risks you can address before they grow into bigger problems.

- *Analyze this funnel and identify the top three areas for conversion improvement.*
- *Suggest three key metrics to track for the upcoming product launch and explain why they matter.*
- *Compare retention rates between two user cohorts and suggest possible causes for differences.*

These prompts aren't just about crunching numbers—they ask AI to explain why the data matters. That shift from raw output to interpreted insight is where the real value lives. Done well, it means fewer surprises, faster pivots, and better-informed bets.

Of course, insight alone is not enough. The way you translate those insights for different audiences often determines whether they drive action or get ignored. This is where AI can help you shift from analysis to alignment, tailoring the same information for executives, teams, and customers so that data not only informs your strategy but also wins the support to execute it.

5. Stakeholder Communication

Even the best insights fall flat if they aren't communicated well. Product managers live in translation mode—adapting the same information for executives, designers, engineers, or customers. AI helps you compress complexity into clear, tailored narratives for each audience.

Stakeholder Communication
Every PM knows the challenge of tailoring the same information for different audiences. AI can help you translate product complexity into concise, audience-specific messaging that resonates.

- *Draft an update for the leadership team summarizing Q2 product performance in 200 words.*
- *Create a one-slide summary of our top 3 roadmap priorities for the next quarter.*
- *Write a customer-facing announcement for our latest feature release in a tone that is informative and engaging.*

These prompts save you from endless rewrites and re-spins. They help you adapt tone, length, and framing without diluting the message. The result: communication that's faster, clearer, and more effective at winning alignment across audiences.

Taken together, these prompt libraries create a repeatable system for accelerating product work. Discovery becomes faster, roadmaps become clearer, design becomes lighter, analysis becomes smarter, and communication becomes easier. When you craft them with precision, AI stops being a novelty and becomes a trusted partner in building products at MACH-10 speed.

Prompts are your leverage; this playbook is how you turn them into impact at MACH-10 speed.

Prompt Recovery Playbook

Even with strong prompts, you won't always nail it on the first try. Treat prompting like iteration: refine until you get exactly what you need.

- **Be more specific:** Add context or clarify the task.

- **Break it down:** Split complex prompts into smaller steps.

- **Refine the format:** Ask for bullet points, tables, or summaries.

- **Iterate:** Try different variants and compare results.

Prompting is a conversation, not a command. The more you interact, the closer AI gets to your intent.

Prompting Isn't Just a Solo Sport

The smartest PM teams treat prompting like product thinking: collaborative, iterative, and worth sharing. Building prompts together makes the whole org more powerful.

Encourage your team to:

- Share effective prompts

- Review outputs together

- Maintain a shared prompt library

When everyone is fluent, the product function benefits from faster insights, fresher ideas, and consistent quality.

> **Mastering prompt engineering is the new literacy for product managers—where the right questions unlock AI's full potential**

Building Prompt Muscle Memory

Like analytics in Chapter 9, prompting gets stronger with repetition. Keep what works, refine it, and store it in a shared space. Treat prompts like reusable code that evolves with your product and market.

Build this kind of prompt muscle memory and your team stops treating AI as a novelty and starts using it as a core system. The benefit isn't just speed but consistency and shared language. As prompts mature into workflows, they stop being tricks and become the base for scalable systems. That is where we turn next: how to codify these practices into playbooks that elevate the product function.

Your Next Step

In the next chapter, we'll explore AI-powered workflows and templates—how to combine tools, prompts, and processes into repeatable systems that scale your impact.

5 Tools to Try This Week	
ChatGPT	For prompt testing and refinement
Notion	For storing prompt libraries
Claude	for structured prompt outputs
coda	For collaborative prompt templates
slack	For team prompt sharing

3 Prompts to Test Today
1. Act as a UX researcher. Summarize these 50 survey responses.
2. List the pros and cons of each option before making a recommendation.
3. Suggest 3 features under 2 weeks of dev time for a fintech app.

Mindset Shift to Adopt

Prompt engineering is not a one-time tactic. It is a capability you build and refine over time. The shift happens when you stop treating prompts as throwaways and start treating them as assets that save time, cut errors, and refine decisions. Mastery comes from iteration: test, tweak, and document what works so your prompts grow with your products and your team's needs.

Reflective Questions

- When do you let tools decide for you instead of treating them as advisors?

- What signals or patterns are you overlooking because you rely too heavily on dashboards?

- How will you train yourself and your team to ask smarter questions as AI evolves?

- What prompts or practices could you standardize now so your team builds consistency instead of relying on luck or one-off wins?

Exercise

Pick one AI system you rely on regularly — analytics, prioritization, or customer insight.

1. Run a scenario with the default AI output.

2. Then re-run it, this time interrogating *why* it gave that answer (ask it to show assumptions, highlight confidence levels, or explain trade-offs).

3. Compare both results and note how your own judgment changes once you see "under the hood."

Key Takeaways

- Vague prompts deliver vague results; precise, context-rich prompts drive clearer insights.

- Strong prompts balance context, clarity, and constraints for outputs that are accurate and actionable.

- Treat prompt engineering as a living discipline that evolves with lessons learned and market shifts.

- Build a prompt library that accelerates speed, alignment, and accuracy across the team.

Chapter 12: Playbooks That Scale

Turning AI Experiments into Repeatable Wins

One enterprise PM noticed her team was running the same onboarding experiment every quarter. They weren't failing, but they were reinventing the wheel. Nothing was codified, so the wins never stuck. By turning the best parts into a playbook of prompts, tools, and decision triggers, she cut the cycle from six weeks to ten days. Her team described it as "stealing from our own best work."

That's the power of codifying workflows. A good workflow makes one team faster. A codified playbook makes the entire organization smarter. The shift is from isolated efficiency to collective momentum. Workflows stop being clever tricks and start becoming the company's operating system.

What Is an AI-Powered Workflow?

At the team level, an AI workflow is a documented, repeatable process. At scale, it becomes an operating system for how the organization works. The distinction is scope: workflows at scale deliver not just speed but governance, enablement, and shared adoption.

Core elements remain the same:

- **AI tools** (e.g., ChatGPT, Figma AI, Mixpanel AI) that generate, analyze, or synthesize information at speed.

- **Prompt templates** that ensure high-quality, consistent inputs.

- **Human checkpoints** where people review, refine, and make strategic decisions.

But scaling workflows adds new layers: documentation in a shared system, cross-team interoperability, and metrics to measure their impact.

When workflows are built this way, they stop being "one person's trick" and become the organization's muscle memory.

> ❝ **Integrating AI into workflows isn't automation alone — it's the art of designing human-centered systems that multiply impact** ❞

Signals from the Edge

One B2B SaaS team I worked with built a workflow for customer escalations. Before, support tickets piled up and engineering jumped in reactively. With a simple AI playbook — clustering tickets, surfacing recurring patterns, and proposing fixes — the team spotted systemic issues faster. Instead of reacting ticket by ticket, they solved the root cause and cut escalations in half. The workflow didn't just save time, it changed the relationship between support and product.

Workflows That Actually Stick

The real value of AI is not in one-off wins but in repeatable workflows that drive consistent impact. The best results come when you apply these workflows to high-stakes, cross-functional processes that happen often enough to shape outcomes. When you embed AI into research, roadmapping, experimentation, and beyond, you turn scattered effort into a system — one that scales clarity, speed, and alignment across your team.

Research Workflow

When research is scattered, insights rarely lead to action. This workflow compresses the cycle from raw data to decision-ready insights:

1. Collect interviews and survey data as a rich base.

2. Use Dovetail AI (or equivalent) to tag and cluster themes.

3. Prompt ChatGPT: "Summarize top 3 user pain points with direct quotes."

4. Share a clean brief with your team for alignment.

This keeps data true to the customer's voice while accelerating delivery.

Roadmapping Workflow

Roadmaps often begin as a messy pile of requests. This workflow filters noise into a clear plan:

1. Aggregate requests and feedback across sources.

2. Use AI scoring in Productboard to rank impact vs. effort.

3. Prompt Claude: "Generate 2 roadmap options: one for growth, one for retention."

4. Align final choice with leadership and engineering.

The result: speed, transparency, and a clear decision trail.

Experimentation Workflow

Experiments should launch quickly and evaluate cleanly. This workflow ensures agility:

1. Identify a friction point in your product.

2. Prompt AI: "Propose 3 A/B test ideas for checkout, with KPIs."

3. Feed winning ideas into the roadmap.

The payoff: tests move from idea to live in hours, not weeks. Each cycle builds momentum, giving teams faster answers and sharper confidence.

Additional Cross-Functional Workflows

At scale, think bigger:

- **Go-to-Market:** Generate persona-specific messaging, check compliance, and deploy into campaigns.

- **Customer Escalation:** Cluster support tickets, propose resolutions, and feed insights to product teams.

- **Executive Decision:** Summarize metrics, model scenarios, and drive faster alignment at the top.

Building Your Own AI Workflows

To move beyond starter workflows and scale across your org:

- Identify repeatable tasks (quarterly planning, release notes).

- Break them into steps with clear outcomes.

- Assign AI tools and prompts to each step.

- Document in a shared hub (Notion, Confluence, Coda).

- Test and refine with real-world use.

At advanced scale:

- Map dependencies across teams.

- Ensure interoperability with systems like Jira and Slack.

- Design checkpoints (peer review vs. compliance sign-off).

- Instrument results to show ROI in time saved and quality improved.

AI-Enhanced Templates

Templates make workflows repeatable at scale. They kill the blank-page problem, enforce consistency, and lock in quality from the start. Instead of reinventing the wheel, teams launch with structures that already capture best practices.

Examples include:

- Research synthesis templates with built-in formatting

- Release note frameworks for consistent detail

- KPI dashboards wired for AI analysis and reporting

At scale, templates evolve into a knowledge design system that's shared, searchable, and version-controlled, much like a Figma library for workflows. The payoff goes beyond efficiency. Templates raise the baseline for everyone, improving quality before refinement even begins. Over time, best practices stop scattering and start compounding into speed, clarity, and quality as the default.

This shift is structural, not cosmetic. It turns everyday work into a system that builds momentum on its own. The next graphic shows how small task-level efficiencies stack into a system that compounds value across the organization.

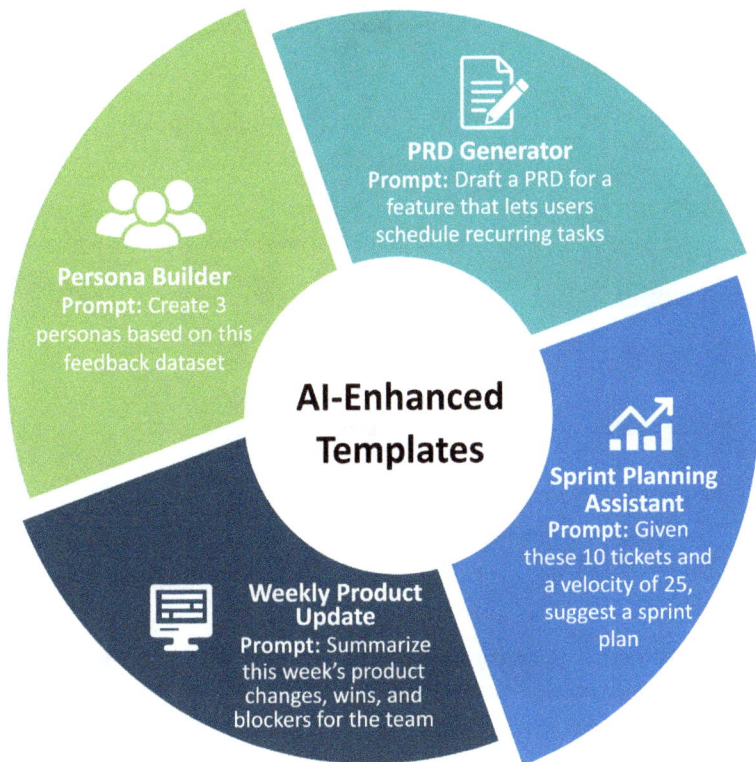

Templates create compounding value — turning small task-level wins into a system of speed, clarity, and quality

Scaling Across Teams

A workflow's value is tied to adoption. Scaling requires more than documentation — it's about culture and enablement:

- **Shared libraries**: build a central hub for prompts and workflows.

- **Onboarding programs**: run training to introduce workflows.

- **AI champions**: nominate a point person per squad.

- **Feedback rituals**: review AI outputs together to normalize usage.

- **Celebration**: highlight wins where AI saved time or improved outcomes.

When adoption spreads, AI workflows become "how we work here," not "something we're trying."

A workflow helps one team. A playbook lifts the whole company.

5 Tools to Try This Week

N Notion	For documenting workflows
coda	For interactive templates
ChatGPT	For workflow generation
productboard	For roadmap workflows
Dovetail	For research workflows

3 Prompts to Test Today

1. Create a workflow for writing release notes using AI.

2. Suggest a repeatable process for sprint planning with AI.

3. Design a research workflow using Dovetail and ChatGPT.

Mindset Shift to Adopt

AI workflows become powerful when they shift from ad hoc fixes to part of the company's nervous system. That takes trust in both speed and quality, and treating documentation as an asset.

Workflows are culture in motion. When AI playbooks become the default, speed and quality become the standard.

Reflective Questions

- Which recurring tasks in your PM role could benefit most from standardized workflows?

- What barriers (technical, cultural, or governance-related) could slow adoption?

- How can you demonstrate ROI to gain leadership sponsorship?

Exercise

Pick a process that spans multiple teams, such as go-to-market planning or quarterly roadmap alignment, and map it into a workflow with standard prompts, clear checkpoints, and measurable outcomes. Share it with cross-functional leaders and refine it together.

Key Takeaways

- AI creates the most value when built into repeatable, standardized workflows.

- Balance automation with human checkpoints to maintain quality.

- Scaling AI is as much cultural as technical, requiring training and shared ownership.

AI won't do your thinking for you. What it does is strip away noise so your judgment can move faster, hit sharper, and carry more weight.

Chapter 13: Proof Over Pitch

Real Results from AI-Driven Product Teams

F rameworks help. Playbooks guide. But, what really changes minds is proof. The case studies ahead show how AI is reshaping discovery, prioritization, and launch. Some are wins worth copying. Others point to what is possible.

Case studies matter because product management runs on judgment under pressure. Frameworks give structure, but stories show it in the wild — real stakes, real constraints, real decisions. They make ideas concrete and turn theory into something you can use. Seeing how others shipped faster, cut risk, or unlocked growth builds conviction. And conviction turns a smart plan into a bold move.

The Anatomy of a Winning Case

Each case study follows the same simple structure so you see not just what was done, but why it worked and how it can be reused:

- **Context:** A quick snapshot of the company, its market, and product.

- **Problem:** The challenge that triggered the need for AI, tied to outcomes not just symptoms.

- **Strategy:** How AI was applied and where it fit into the workflow.

- **Tools:** The platforms and systems used to make it work.

- **Results:** What changed, what the team learned, and how it created momentum.

This structure turns case studies into practical assets. Each AI experiment becomes a playbook your team can adapt and scale. Over time, these examples build into a library of proven practices that accelerate adoption and give leaders the confidence to push faster.

> ❝ **You can't improve what you don't measure—AI's promise becomes real when success metrics are clear, aligned, and actionable** ❞

Case Study 1 – Healthtech Startup

In healthcare, speed and clarity are essential. This startup used AI to turn a flood of feedback into clear signals for faster, better decisions. Instead of guessing from scattered inputs, they focused only on the patterns AI surfaced as most urgent, which kept the team aligned and moving.

Case Study 1: Accelerating Discovery at a Healthtech Startup

Context: A Series A healthtech startup building a mobile app for chronic condition management.

Problem: The PM team struggled to synthesize feedback from thousands of users across support tickets, app reviews, and surveys.

AI Strategy:
- *Used Dovetail AI to tag and cluster feedback*
- *Prompted ChatGPT to summarize top pain points and feature requests*
- *Generated personas and opportunity areas in hours, not weeks*

Outcome:
- *Reduced research synthesis time by 80%*
- *Identified a new feature opportunity that increased retention by 12%*

Lesson: AI can turn noisy feedback into clear direction—fast.

Lesson: In data-heavy industries, AI turns overload into clarity. The PM mined existing data for insights and turned noise into advantage, proving speed comes from focus, not more data. By cutting through the clutter, they made decisions faster, aligned the team, and turned scattered signals into a clear path forward.

Case Study 2 – B2B SaaS Company

A SaaS team kept losing RFPs. Their decks looked polished but missed key patterns. AI surfaced them in hours, pulling the recurring asks buried in hundreds of pages. The next pitch hit what mattered, spoke the buyer's language, and closed deals that had been slipping away for months.

Case Study 2: Smarter Roadmapping at a B2B SaaS Company

Context: A mid-sized SaaS company serving enterprise clients.

Problem: Roadmap decisions were slow and politically charged, with unclear prioritization criteria.

AI Strategy:
- *Used Productboard AI to score features based on feedback and strategic fit*
- *Prompted Claude to generate roadmap scenarios under different constraints*
- *Shared AI-generated trade-off summaries with stakeholders*

Outcome:
- *Cut roadmap planning time by 50%*
- *Improved stakeholder alignment and reduced pushback*

Lesson: AI can depoliticize prioritization and clarify trade-offs.

Lesson: In complex sales, AI pulls hidden signals from dense documents. Proof beats polish. MACH-10 PMs know that winning big deals is not about flash or flawless design decks. It is about showing buyers you understand what matters most to them. Dense RFPs, compliance requests, and contract redlines often hide those signals in walls of text. AI cuts through the noise, clustering demands and surfacing patterns humans miss. That turns guesswork into targeted strategy and makes your pitch feel inevitable instead of hopeful.

Case Study 3 – Fintech App

A Fintech app was drowning in feature requests. Everyone wanted new rewards and gamified perks, and the roadmap leaned into them. But growth had flatlined. When the team ran their support logs and user reviews through AI, a pattern appeared: reliability complaints dwarfed everything else. Crashes during transfers, failed deposits, and inconsistent balances were eroding trust. The team stopped chasing shiny features and rebuilt the transaction engine. Once reliability stabilized, support tickets dropped, user trust surged, and growth returned without a single new perk launched.

Case Study 3: Scaling Growth Experiments at a Fintech App

Context: A fast-growing fintech app focused on budgeting and savings.

Problem: The growth team struggled to design and analyze A/B tests quickly enough to keep pace with user growth.

AI Strategy:
- *Used GrowthBook for feature flagging and experimentation*
- *Prompted ChatGPT to generate test ideas and success metrics*
- *Used Mixpanel AI to analyze results and recommend next steps*

Outcome:
- *Increased experiment velocity by 3x*
- *Unlocked a new onboarding flow that boosted activation by 18%*

Lesson: AI can make growth experimentation faster, smarter, and more scalable.

Lesson: In noisy markets, AI surfaces the problems that matter. MACH-10 PMs ignore praise and chase what drives retention and revenue. Positive feedback can mislead, while churn often hides in the corners. AI cuts through bias to show what users actually need fixed.

Case Study 4 – Consumer App

A consumer app had glowing reviews but slipping retention. AI flagged login friction buried in one-star comments. Password resets failed and two-factor prompts timed out. Fixing onboarding cut churn in half, retention rebounded, and praise finally matched the experience.

Case Study 4: AI-Augmented Delivery at a Consumer App

Context: A consumer app with weekly release cycles and a lean engineering team.

Problem: PMs spent too much time writing release notes, grooming backlogs, and coordinating QA.

AI Strategy:
- *Used Linear AI to summarize issues and suggest sprint plans*
- *Prompted Claude to generate release notes from commits and tickets*
- *Used Testim to automate regression testing*

Outcome:
- *Saved 6+ hours per week per PM*
- *Reduced post-release bugs by 30%*

Lesson: AI can streamline delivery and free up PMs for strategic work

Lesson: AI finds signal in the dark corners. MACH-10 PMs mine frustrations to stop growth leaks.

These case studies prove that AI in product management is not a future trend. It is an operational reality. The teams getting the most from AI aren't dabbling. They are embedding it into core processes, measuring results, and refining with every cycle.

Proof wins attention. ROI keeps it.

In the next chapter, we'll build a framework for tracking ROI, speed, and strategic lift — because in product management, proof may open the door, but measurable impact keeps it open.

5 Tools to Try This Week	
Dovetail	For feedback synthesis
productboard	For prioritization
GrowthBook	For experimentation
Linear	For delivery planning
Claude	For summarizing case studies

3 Prompts to Test Today

1. Summarize top user complaints from 1,000 support tickets.
2. Generate 3 roadmap scenarios based on these constraints.
3. Write a case study summary for a successful AI integration.

Mindset Shift to Adopt

AI in product management has moved from "emerging" to "expected." The companies moving fastest are building advantages that will be difficult for slower adopters to catch. By studying real-world applications, you shortcut the trial-and-error phase and focus your efforts on proven, high-impact use cases.

Reflective Questions

- How deeply is AI integrated into my daily product workflows today?

- Which processes would benefit most from AI-driven speed or clarity?

- Where am I still relying on manual work that slows down decisions?

- How often do we document and share successful AI use cases across the team?

- What is preventing our AI adoption from moving from experiments to consistent practice?

Exercise

Choose one recent or ongoing AI integration in your work — whether small (like AI-assisted release notes) or large (like AI-driven roadmap prioritization). Document it using the case study template provided in this chapter. Be sure to include the **context, problem, approach, results, and lessons learned**. Then, share it with your team for feedback. This does two things: it validates your approach and creates a reusable artifact your organization can refine and scale.

Key Takeaways

- **Define** the problem before choosing tools and prompts.

- **Embed** AI into existing workflows to create efficiency without overhaul.

- **Oversee** outputs to keep them aligned with strategy and brand.

- **Measure** results to build buy-in and accelerate adoption.

- **Share** case studies to turn wins into repeatable practices.

- **Stack** small wins until they compound into lasting advantage.

Chapter 14: ROI at MACH-10

Measure Speed, Impact, and, Real Value

I n the previous chapter, we saw how real-world examples bring AI-powered product management to life. But even the most innovative applications lose momentum if you can't prove their value. Without measurement, AI adoption is just guesswork. To secure buy-in, maintain trust, and make informed calls, PMs must define success metrics, track performance, and communicate results in ways that resonate.

This chapter gives you a framework to measure productivity gains, decision quality, and business outcomes. You'll learn to set baselines, monitor progress, and tell a compelling story about AI's value to your product and team. Treat AI as a measurable investment, not a novelty, and it will continue to deliver strategic returns over time.

Without Metrics, It's Just Hype

Integrating AI without measurement risks wasted resources, unclear benefits, and fading support. Clear metrics let teams and leaders track progress, celebrate wins, and spot underperformance. The result: alignment, accountability, and demonstrable ROI.

Beyond Hype: Proving AI's ROI

AI delivers value in multiple ways, but three categories consistently matter for PMs: **efficiency and productivity**, **decision quality and accuracy**, and **user or business impact**. Clear metrics turn AI from a buzzword into a measurable driver of performance, enabling PMs to demonstrate real impact, guide smarter decisions, and align teams around tangible outcomes.

Below, we'll break each category down and give you concrete examples to measure them effectively.

1. Efficiency and Productivity Metrics

These metrics show whether AI is truly lifting the execution load or just adding another layer of noise. The goal is simple: free PMs from repetitive tasks so they can focus on strategy and judgment.

- **Time Saved** – Quantifies the reduction in hours spent on routine work.
  Example: Weekly research synthesis time drops from 10 hours to 3 after using AI to cluster and summarize interview data.

- **Task Automation %** – Tracks the share of repetitive tasks that no longer require human effort.
  Example: 70% of release notes auto-generated with AI, freeing PMs and engineers to focus on customer-facing improvements.

- **Cycle Time Compression** – Measures how AI accelerates key workflows from start to finish.
  Example: Backlog grooming cycles shrink from a week to a day with AI triaging duplicates and clustering themes.

2. Decision Quality and Accuracy Metrics

AI should sharpen your choices, not just speed them up. These metrics show whether decisions are more reliable and defensible with AI in the loop—steering with clearer signals instead of louder opinions.

- **Prediction Accuracy** – How often AI forecasts prove correct against real outcomes.
  Example: Sprint forecasting accuracy improves from 60% to 85% with AI-driven velocity models.

- **Error Reduction Rate** – The drop in mistakes or defects when AI augments decision-making.
  Example: QA defect rates fall 30% after automated test generation catches edge scenarios humans missed.

- **Decision Confidence Scores** – How often teams feel they have enough evidence to commit.
  Example: Stakeholder survey shows a 40% rise in confidence when roadmap trade-offs are backed by AI scenario modeling.

Sharper decisions strengthen the product team, but the real test is how they land with customers and the business. That's where impact metrics take center stage.

3. User and Business Impact Metrics

At the end of the day, AI must move the needle where it matters most: customer experience and business growth.

- **User Satisfaction (NPS/CSAT)** – Measures shifts in user sentiment tied directly to AI features or faster feedback loops.
  Example: NPS jumps from 42 to 56 after an AI-powered onboarding flow reduces friction for new users.

- **Revenue Impact** – Quantifies incremental revenue attributed to AI capabilities.
  Example: Personalized recommendations drive a 12% uplift in average order value over three months.

- **Retention & Churn Metrics** – Shows whether AI reduces attrition or extends user lifetime value.
  Example: AI-driven churn prediction cuts monthly churn from 5% to 3% by enabling proactive retention campaigns.

- **Operational Cost Savings** – Tallies direct savings from automation or improved efficiency.
  Example: Automating customer support triage saves $200K annually in staffing costs without hurting satisfaction.

Measuring AI That Matters

Collecting metrics is only half the battle. A clear framework ensures you measure the right things the right way:

Identify objectives for each AI initiative (productivity gains, accuracy, efficiency).

1. Select KPIs that align with those objectives.

2. Establish baselines before implementation.

3. Use dashboards and analytics to track KPIs in real time.

4. Review results regularly and refine AI usage.

This keeps you from chasing vanity metrics and anchors impact to tangible business results.

> " **Scaling AI is less about technology and more about culture—creating ecosystems where humans and machines thrive together** "

Turning Numbers into Narrative

Effective PMs do more than implement AI. They make its impact visible and credible to stakeholders. The way you communicate results determines whether AI feels like a side experiment or a core driver of business performance.

Strong communication often takes three forms:

- **AI Impact Dashboards**: Monthly or quarterly views that show measurable gains in efficiency, speed, or outcomes. These highlight trends over time, not just one-off wins.

- **Story-Driven Summaries**: Short narratives that bring the numbers to life with case studies, user quotes, or team testimonials. Data tells the "what," but stories explain the "why it matters."

- **Clear Attribution**: Direct links between performance shifts and AI implementations, so leaders can see the cause and effect instead of chalking results up to coincidence.

The best PMs go further by tailoring the message to the audience. Executives want the business impact, designers want the workflow impact, and engineers want the technical clarity. By translating the same outcomes through different lenses, you turn raw results into alignment and momentum.

When you make AI's value clear, adoption accelerates. Teams stop viewing it as an optional add-on and start treating it as part of the operating system of product management.

Signals from the Edge

At a health tech startup, the PM began attaching short AI-generated summaries to each release. Instead of dense changelogs, executives received two clear paragraphs that explained the trade-offs, metrics, and outcomes in plain language. The summaries connected the numbers to the story behind them, making AI's impact easy to see. Within two quarters, executive alignment sped up noticeably, and adoption of AI-driven reporting spread beyond product into operations. Trust wasn't just built—it became momentum.

If you can measure it, you can multiply it.

5 Tools to Try This Week	
ChatGPT	For summarizing research and generating ideas
Notion	For writing product briefs and meeting notes
Figma	For early design exploration
Jira	For backlog insights
Amplitude	For product usage analytics

3 Prompts to Test Today
1. What are 5 ways AI can improve my current product workflows?
2. Summarize the key benefits of AI for PMs in under 100 words.
3. List AI tools that align with my product goals.

Mindset Shift to Adopt

Stop treating AI as an experiment. Treat it as a teammate you evaluate and hold accountable. Measurement is not bureaucracy; it gives AI a real seat at the product table. Apply the same rigor to its outputs as you do to human work: set expectations, track performance, and refine results. Consistent measurement builds trust and proves AI is a reliable driver of strategy and execution.

Reflective Questions

- Which success metrics best align with your current AI implementations?

- Where do you lack baselines today, and how can you establish them quickly?

- How effectively are you communicating AI's value to stakeholders?

- If asked tomorrow, could you quantify the ROI of your AI investments?

Exercise

Pick one active AI project. Build a simple dashboard (Notion, Looker, or Sheets) to track performance against clear metrics. Share it in your next stakeholder update, gather feedback, and refine until the dashboard becomes a trusted reference point.

Key Takeaways

- Clear, measurable goals are critical to demonstrating AI's tangible value.

- Metrics should cover productivity, decision quality, and user/business impact.

- Treating AI like a teammate fosters ongoing improvement and accountability.

Chapter 15: Trust is the Moat

Leading with Responsibility in the Age of AI

T he most powerful AI capabilities mean nothing if they erode trust. Ethics is not an add-on; it is a core responsibility. As AI shapes decisions and experiences, PMs must address risks around bias, transparency, privacy, and accountability.

An eCommerce PM faced pressure to launch a recommendation engine that boosted revenue but also surfaced high-return products. Flagging the risk and refusing to ship avoided churn and built long-term loyalty.

Speed and impact mean little if they come at the cost of fairness or credibility. Responsible AI protects users, strengthens your brand, and creates products that last.

Reflective Questions

- Where in your product could speed or optimization be hiding risk?

- What signals would tell you trust is in danger before it is lost?

> **Ethics isn't a constraint—it's the compass guiding AI innovation toward trust, fairness, and lasting value**

Trust Is the Moat

PMs sit at the intersection of users, data, and technology. With AI-driven tools making more decisions than ever, the ethical stakes have never been higher. A single misstep can do more than dent a quarterly report—it can permanently erode trust, harm users, or invite regulatory scrutiny.

Responsible AI means being intentional at every stage of the lifecycle. It's about anticipating risks before they escalate and embedding safeguards so fairness, transparency, and respect for privacy are baked in from the start. These principles cannot be treated as nice-to-haves. They must become the default operating system of product management in the AI era.

The Trust Traps: Core Ethical Challenges

When integrating AI into products, three pressure points surface again and again. Each requires both awareness and a plan of action.

1. Bias and Fairness

AI models reflect the data they're trained on. If that data contains bias, the outputs will too, often reinforcing harmful stereotypes or unequal outcomes. Left unchecked, bias can show up in product recommendations, credit approvals, or hiring filters, damaging trust and creating real harm.

Examples of Bias:

- Gender or racial bias in hiring algorithms

- Search rankings or recommendations that exclude minority groups

- Customer service bots treating dialects or languages inconsistently

Mitigation Tactics:

- Conduct routine audits on both datasets and model outputs

- Build diverse, multidisciplinary teams for training and validation

- Stress-test models against edge cases, not just averages

- Communicate openly with users about how AI-driven decisions are made

Bias doesn't live only in code. It lives in the assumptions, the context, and the shortcuts teams take under pressure. Addressing it takes both technical rigor and a cultural commitment to fairness.

2. Data Privacy and User Trust

AI thrives on data, but more is not always better. Collecting information without a clear purpose increases exposure to risk and weakens user trust. Mishandled data can erase years of goodwill in a single incident.

Privacy Risks:

- Unauthorized use of personal information

- Security breaches or leaks of sensitive data

- Overcollection of unnecessary information that adds liability without value

Best Practices:

- Limit collection to only what is necessary (data minimization)

- Be transparent: tell users what you collect, why, and how it's secured

- Embed Privacy-by-Design into products so protections are engineered in from day one

- Run regular privacy reviews as part of your product development cycle

Privacy isn't just a legal requirement. It's a brand promise. Handle data poorly and you lose more than compliance—you lose trust, loyalty, and long-term market advantage.

3. Explainability and Transparency

If you can't explain how your AI made a decision, you can't own its outcomes. Opaque systems create doubt, weaken adoption, and expose you to scrutiny you can't answer.

Explainability Issues:

- Difficulty defending automated decisions to users or regulators

- Inability to troubleshoot errors or unexpected outcomes

- Internal uncertainty over accountability when results are unclear

Enhancing Transparency:

- Use explainable AI (XAI) or interpretable models whenever possible

- Provide plain-language explanations for important AI-driven decisions

- Publish internal or external transparency reports that clarify design choices and oversight processes

- Train teams on how to communicate AI's role to stakeholders and customers

Transparency shifts AI from mysterious to understandable, driving adoption, accountability, and resilience to scrutiny. These challenges are interconnected, and the matrix below maps them by risk and opportunity to show PMs where to act boldly, tread carefully, and monitor closely

Regulatory & Ethical Risk Matrix
Balancing Risk and Opportunity in AI PM

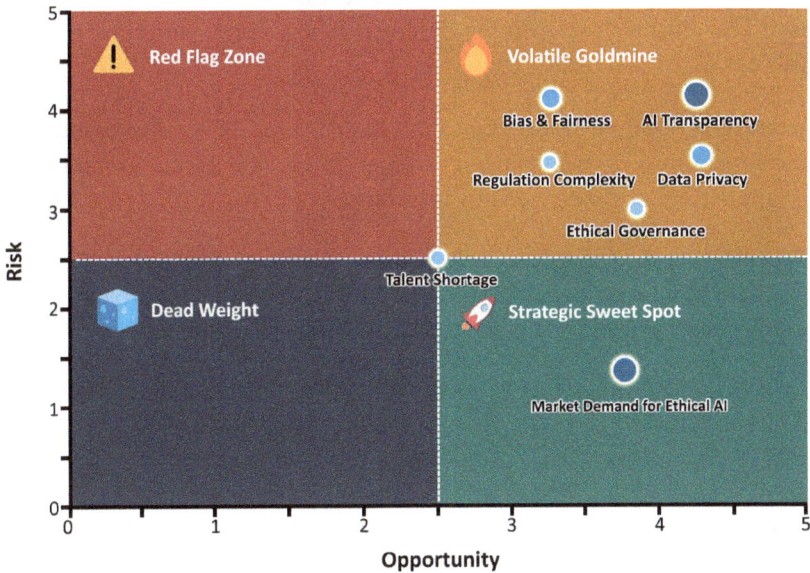

Risk = regulatory/ethical volatility | Opportunity = value creation potential

Balancing ethical risk and strategic value in AI products.

The Seven Pillars of Responsible AI

Responsible AI doesn't happen by accident. You need guardrails that hold, even under pressure. Think of these seven pillars as the frame of your product house. Leave one out, and the whole thing leans. Build them all in, and you get strength you can trust at scale.

1. Bias & Fairness AI models reflect the data they're trained on. This pillar ensures that you proactively audit data and models to prevent reinforcing harmful stereotypes or unequal outcomes, ensuring the product serves all users equitably.

2. Explainability & Transparency If people can't see how decisions are made, they won't trust the outcome. This pillar focuses on making AI decisions interpretable and communicating the "why" behind your choices—internally and externally. Opacity breeds suspicion.

3. Governance Who owns what? Who gets the final call when risk shows up? Clear roles, escalation paths, and accountability lines keep ethics from falling into the cracks.

4. Accountability No black boxes. If AI makes a decision, someone human still owns it. This pillar is about defining clear ownership so problems aren't shrugged off as "the algorithm's fault."

5. Inclusivity Homogenous teams miss blind spots. This pillar ensures that diverse perspectives are included in the design and review process to expose bias and shape better, more representative systems.

6. Data Privacy AI thrives on data, but trust requires stewardship. This pillar covers the critical practices of data minimization, security, and providing users with clear control over their information.

7. Continuous Monitoring AI drifts. Models age. Assumptions break. This pillar establishes the need for regular audits and performance checks to ensure yesterday's decision logic doesn't become tomorrow's liability.

Put these seven together, and you don't just avoid disasters—you send a signal. To your team. To your customers. To the market. You're not just moving fast. You're moving fast with guardrails that make speed sustainable.

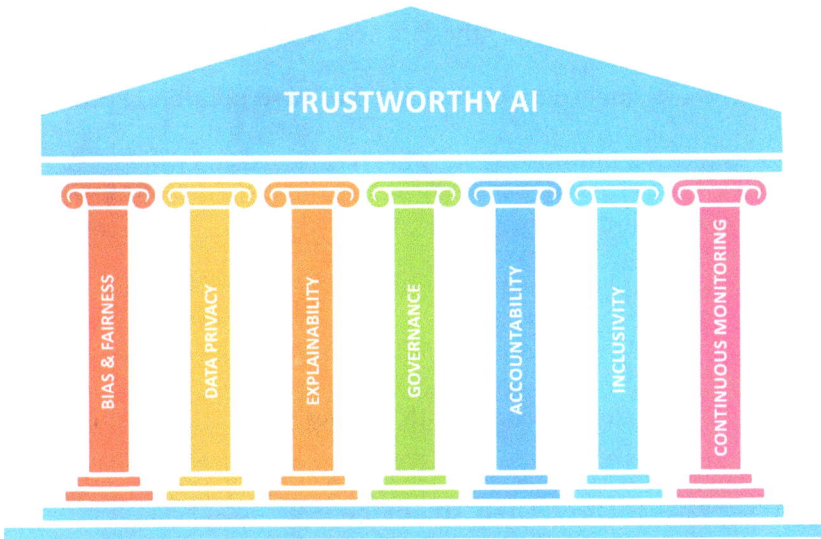

Ethical strength starts with structure

AI Ethics Checklist

Use this list during sprint reviews or release planning to spot gaps and assign next steps.

Bias and Fairness

- Have we tested outputs for potential bias across gender, race, or other sensitive attributes?

- Is the training data diverse and representative of real users?

- Who reviewed this for unintended consequences?

Transparency and Explainability

- Can we explain how this model made its decision in plain language?

- Do we provide users with a way to understand or challenge automated outcomes?

- Is documentation clear and accessible internally?

Privacy and Data Use

- Are we collecting only the data necessary for this feature?

- Have we applied data minimization and secure storage practices?

- Is the user informed about what is collected and why?

Accountability and Oversight

- Who is explicitly responsible if this AI system fails or misbehaves?

- Do we have monitoring in place for unexpected outputs or drift?

- Is there a clear escalation path if issues arise?

Putting the Pillars Into Practice

The value of these pillars is that they give product managers a lens for trade-offs. AI decisions often feel like choosing between speed and safety, personalization and privacy, or efficiency and fairness. By grounding those choices in a set of non-negotiable principles, you avoid slipping into short-term wins that create long-term risks.

Each pillar also reinforces the others. Governance without transparency feels bureaucratic. Transparency without accountability leads to finger-pointing. Inclusivity without monitoring risks becoming performative. But together, they create a system where ethical standards are not bolted on at the end; they are baked into how the team thinks, decides, and delivers.

This is where ethical AI moves from a compliance checkbox to a competitive advantage. Products that are fair, explainable, and trustworthy do more than avoid harm; they stand out in crowded markets by signaling credibility and long-term reliability. Clear ethics also build confidence internally, empowering teams to innovate faster without second-guessing the integrity of their decisions.

From Principles to Daily Discipline

Principles on their own can feel like posters on a wall—aspirational, but not always actionable. What makes them powerful is their translation into day-to-day behaviors and checks that guide real product work. For PMs, that means having a way to pressure-test backlog items, roadmap decisions, and design reviews against ethical standards before they go live.

The table that follows bridges that gap. It turns the pillars of responsible AI into a **working toolkit** you can bring into sprint planning, stakeholder reviews, or even regulatory conversations. Think of it less as a static checklist and more as a **living playbook**—something you revisit, refine, and evolve as your product grows and as regulations, customer expectations, and risks shift.

By operationalizing ethics this way, you move from good intentions to measurable practices. You're not just *hoping* your product is responsible—you're building the guardrails that make responsibility inevitable.

AI without trust is a product without a future.

Pillar	Actions
Bias & Fairness	Audit datasets and models for bias. Apply fairness metrics and mitigation strategies regularly
Data Privacy	Protect user data with anonymization, encryption, and strict access controls
Explainability	Provide clear, understandable rationales for AI outputs using interpretable models or explainers
Governance	Establish clear policies, oversight boards, and guidelines for AI ethics
Accountability	Assign clear responsibilities for AI performance and review regularly
Inclusivity	Prioritize diversity and inclusion in data, models, and product teams
Continuous Monitoring	Continuously audit AI outputs for unintended consequences and drift

Principles are nothing without follow-through

Mindset Shift to Adopt

Ethics isn't a constraint. It's a compass. Weaving fairness, transparency, and privacy into your workflow builds products that inspire confidence, not suspicion. The speed and impact of AI matter little if you can't defend how you got there. Responsible PMs treat ethics as fuel for sustainable innovation. Ethics protects you from shortcuts that damage credibility. It gives your team the license to move faster because stakeholders believe in how you operate.

Reflective Questions

- What ethical risks might your current AI use cases create?

- How transparently does your organization communicate AI-driven decisions?

- Who is accountable when AI gets it wrong, and is that accountability clear?

Exercise

Create an AI Ethics Checklist covering bias, transparency, privacy, and accountability. Use it during your next sprint retrospective to spot gaps, assign action items, and capture learnings. Over time, refine it with team input and make it a standard part of every release cycle so ethical checks are treated with the same rigor as QA or security reviews.

Exercises like these make ethics real by forcing teams to look at how AI plays out in their own workflows. But good intentions fade fast without structure. That's why many high-performing teams adopt a simple ethics checklist—something concrete they can run through at the end of a sprint or before a release. It shifts the conversation from "we'll think about ethics later" to "let's confirm we've covered the essentials right now."

Key Takeaways:

- Ethical product management is essential for the sustainable use of AI.

- Bias, privacy, and explainability require ongoing management, not one-time fixes.

- Clear frameworks and continuous oversight help ensure responsible AI at scale.

- Ethical AI practices can differentiate your product, deepen trust, and safeguard your brand

Chapter 16: What's Next Is Already Here

Navigating the Frontier of Product Leadership

In the previous chapter, we explored how ethical considerations form the backbone of sustainable AI adoption. Responsible use of AI is not just about preventing harm; it's about building trust that makes bold innovation possible. That trust will be even more essential as we step into an AI-native era of product management—one where AI is not just a tool in the PM toolkit, but the underlying infrastructure shaping how we build, decide, and lead.

AI is more than another technology trend. It is a tectonic shift that will transform not only what we deliver, but how we work, how we collaborate, and how we define success. For Product Managers, this is both a challenge and an invitation: to evolve, to lead, and to architect the future of our craft.

The PM Role Is Evolving—Again

The PM role has always adapted to change, but AI is accelerating that evolution at an unprecedented pace. The shift is not about replacing PMs—it's about amplifying their strategic influence. By offloading repetitive execution to intelligent systems, PMs can focus on shaping vision, designing systems, and guiding high-stakes decision-making.

This transformation can be understood through three core shifts:

From Generalist to Strategist
PMs will spend less time executing routine tasks and more time clarifying vision, aligning stakeholders, and building scalable systems.

From Facilitator to Architect
PMs will go beyond coordinating teams to designing the AI-powered workflows, prompts, and human–machine systems that bring products to life.

From Decision-Maker to Decision-Orchestrator

PMs will use AI to simulate, model, and stress-test options—then guide teams through the best path forward.

The Evolution of the Product Manager Role in the AI Era

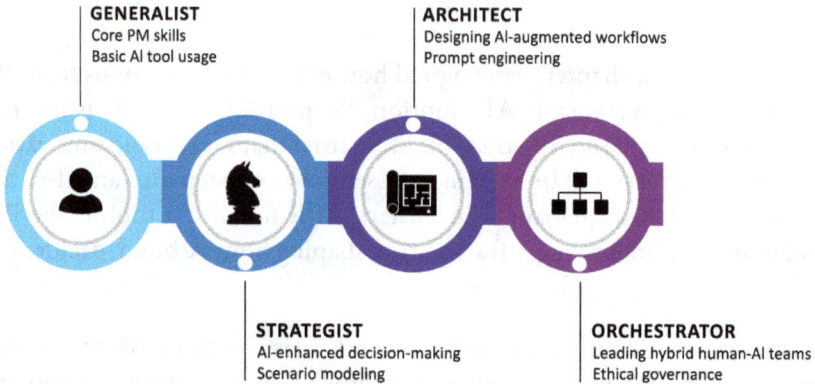

GENERALIST
Core PM skills
Basic AI tool usage

ARCHITECT
Designing AI-augmented workflows
Prompt engineering

STRATEGIST
AI-enhanced decision-making
Scenario modeling

ORCHESTRATOR
Leading hybrid human-AI teams
Ethical governance

The future arrives early for those who act early

The AI-Native PM Skillset

AI is not just changing the PM role—it's redefining the skills required to excel. In this new environment, PMs will need to combine traditional leadership strengths with a fluency in AI's capabilities, limitations, and strategic applications.

These are the skills that will separate high-impact AI-native PMs from the rest:

- **Prompt Crafting** – Writing precise, context-rich prompts that consistently generate high-quality AI outputs.

- **AI Literacy** – Understanding the strengths, weaknesses, and risks of various AI models.

- **System Design** – Building scalable processes that integrate AI efficiency with human judgment.

- **Ethical Foresight** – Proactively identifying bias, privacy, and fairness considerations before they become issues.

- **Narrative Thinking** – Translating AI-driven insights into stories that inspire action and alignment.

When I first tested GPT-4, I asked for a "detailed product spec." What I got back was 1,200 words of filler—technically correct, practically useless. The fix wasn't the tool. It was me learning how to frame the ask. That lesson hit harder than any article or webinar: prompting is briefing, and briefing is everything.

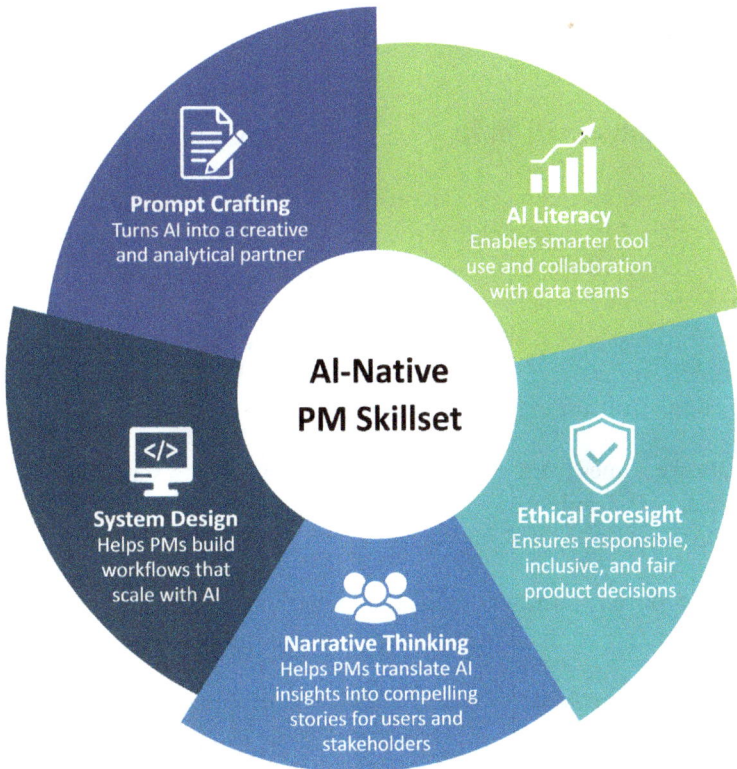

Build the stack: prompts, literacy, ethics, narrative, systems

Future Team Structures

AI will also reshape the structure of product organizations. Teams will become leaner, faster, and more adaptive. What once required a room of specialists will now be achievable with smaller squads augmented by AI.

The future may include:

- Autonomous squads powered by AI-augmented workflows

- Hybrid roles such as Prompt Engineer, AI Product Strategist, and UX–AI Researcher

- Cross-functional AI councils overseeing governance, ethics, and experimentation

For PMs, leading these teams will demand balance: providing alignment without suffocating autonomy, enabling speed without losing sight of responsibility.

Why Humans Still Win

As AI expands its reach, the human qualities that define great PMs will only grow in importance. AI can crunch numbers, run predictions, and strip away operational noise, but only humans can breathe meaning into the results. Without that human filter, you risk falling into the trap of letting the machine decide what matters. And history has shown us how poorly that can play out. Just think of HAL 9000 from *2001: A Space Odyssey* deciding who got to keep breathing on the spaceship. Brilliant logic, catastrophic judgment.

The real edge comes from elevating machine outputs into something resonant, contextual, and deeply valuable. That is where the core strengths of product leadership shine:

- **Empathy**: Seeing beyond the metrics to understand the unspoken needs, emotions, and contexts that shape human behavior.

- **Judgment**: Making nuanced trade-offs in ambiguous situations where no dataset provides certainty.

- **Creativity**: Bridging insights across domains to generate solutions that surprise and delight.

- **Leadership**: Inspiring people, navigating resistance, and aligning stakeholders toward a shared vision.

AI will do the heavy lifting, but PMs will continue to do the meaning-making. The best product leaders will treat AI like a capable partner, not a replacement. Machines can hand you the raw material, but it takes human insight to craft it into strategy, vision, and progress that people actually believe in.

AI may write the draft, but only you can write the story. Machines can generate options, but it takes a human to know which one belongs in the roadmap.

AI-Native PM Skillset Radar Chart

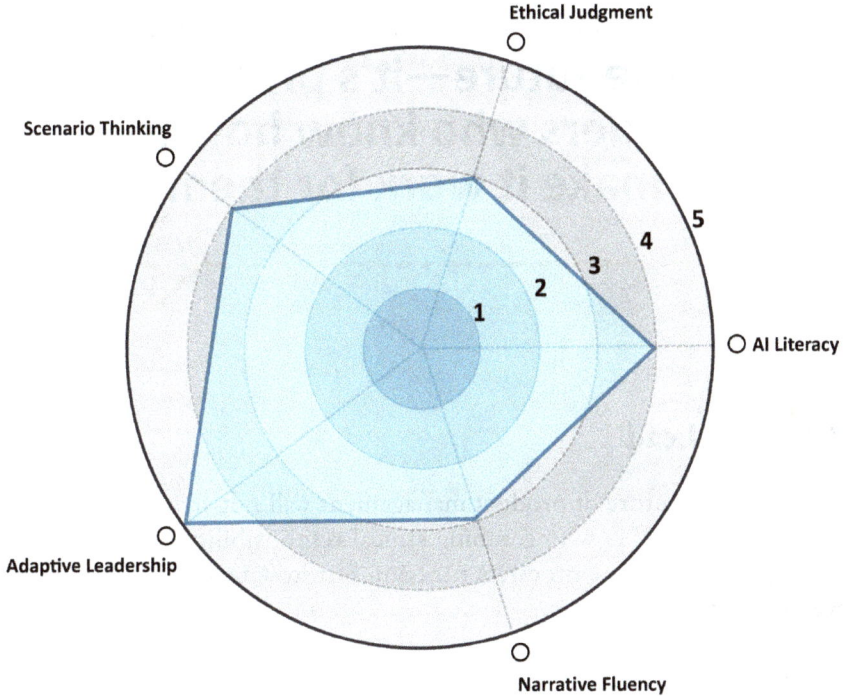

Skill radar for the AI-native PM: see your strengths at a glance

What Won't Change

While methods and mechanics evolve, the essence of product management remains constant. The PM's north star is not efficiency. It is impact.

The enduring responsibilities are:

- Deeply understanding users and their problems

- Aligning teams around a shared vision

- Delivering meaningful value, not just features

AI may change the engine that powers the work, but the destination remains the same: better products, happier users, stronger businesses.

> **"It's not AI that will build the future—it's product leaders who know how to make it work for them"**

A Call to Lead

The AI-native future of product management will not arrive on its own. It will be built by PMs who combine ethical responsibility with AI fluency, shaping not only how products function but how teams, companies, and even industries operate.

This means:

- Understanding the user journey at a deeper level than ever before

- Bridging strategy and execution with AI as a partner

- Navigating seamlessly between technical, design, and business disciplines

- Thinking in systems, and increasingly, thinking in AI systems

A Day in the Life of a PM (2030)

7:45 AM: Coffee in hand, you open your inbox. Hundreds of emails, Jira alerts, and Slack pings. Somewhere in the noise are real insights, but no time to find them.

8:15 AM: In design review, three user flows are debated. Last week's usability data is buried. The meeting spins. Another follow-up is booked.

10:00 AM: Growth review drags. Everyone has a gut feeling, but no reliable numbers. Two hours gone. No decision made.

12:00 PM: A bug report flags possible bias. You log it and ping engineering. Weeks later, nothing's fixed, but the questions start coming your way.

2:00 PM: The afternoon vanishes into slide-building hell. Copy, paste, format, repeat. The customer call you planned? Postponed again.

5:30 PM: Laptop closed, shoulders tight, energy gone. The backlog is longer than this morning. Tomorrow will be worse.

A Day in the Life of a PM (2030)	A Day in the Life of an AI-Native PM (2030)
Drowning in emails, Jira, and Slack noise	AI curates insights, flags risks, strips noise
Debates drag on, data buried in spreadsheets	AI generates flows, creates tickets, team leaves aligned
Endless arguments, no decisions	AI runs simulations, top scenarios greenlit fast
Bias issue logged, no action for weeks	Bias alert caught, fixed in under an hour
Hours wasted on slide formatting	AI builds exec update, you spend time with customers
Exhausted, progress feels incremental	Clear, energized, leading with focus and strategy

The shift from manual chaos to AI-powered clarity in a PM's workday

A Day in the Life of an AI-Native PM (2030)

7:45 AM: Coffee in hand, you open your dashboard. Overnight, AI curated key insights, flagged risks, and cleared the noise. You start with clarity.

8:15 AM: In design review, AI generates three user flows from usability data. The best option is clear. Jira tickets are drafted before the meeting ends.

10:00 AM: In growth review, AI models five pricing scenarios with forecasts and sentiment. The team aligns and greenlights two in 20 minutes.

12:00 PM: A bias alert pops up. You and your ethics lead audit, fix, and close the issue in under an hour.

2:00 PM: Instead of slides, AI builds a clean, exec-ready update. You spend that hour with a customer, gathering insights dashboards can't show.

5:30 PM: You wrap focused and energized. AI handles the load so you can lead with strategy, empathy, and creativity.

5 Tools to Try This Week	
ChatGPT	For strategic scenario modeling and prompt experimentation
Notion	For capturing evolving team workflows and AI learning goals
miro	For visualizing future team structures and product strategies
Claude	For generating ethical AI frameworks and leadership communication
slack	For facilitating AI-powered team alignment and collaboration

3 Prompts to Test Today
1. What are the top 3 skills PMs will need in an AI-native world?
2. Design a future team structure for AI-augmented product development.
3. Identify key ethical risks and opportunities for AI in product leadership.

Mindset Shift to Adopt

AI takes on the heavy lifting: the noise, repetition, and churn that do not need a human touch. The PM's value is in what machines cannot do: leading with empathy, guiding with clarity, and making meaning from context. The future is not about speed, but about asking sharper questions, framing smarter problems, and tying insights to strategy while AI handles the rest.

Reflective Questions

- Which future skills and mindsets feel most critical for you personally?

- How will your product approach change in the next 12 months based on what you've learned?

- What steps can you take now to integrate AI into your workflow with intention and purpose?

Exercise

Write your personal "AI-Powered PM Roadmap" for the next year:

- Identify three skills to develop

- Define three AI integrations to implement

- Set three measurable goals for growth

Key Takeaways

- AI is reshaping the PM role — from generalist to strategist, facilitator to architect, and decision-maker to orchestrator.

- Tomorrow's PMs will master prompts, AI literacy, system design, ethical foresight, and the art of turning data into compelling stories.

- Organizations will evolve — leaner teams, new hybrid AI roles, and formal governance frameworks.

- The human edge — empathy, judgment, creativity, leadership — will only grow in importance.

- PMs fluent in AI and ethics will lead the next wave of innovation with trust, clarity, and adaptability.

Chapter 17: Leading at MACH-10

Locking in The Multiplier Mindset

In Chapter 16, we explored how AI is reshaping the product management role, transforming PMs into strategists, architects, and orchestrators of decisions. Now it's time to push further into the future. It's not just about tools or techniques. We are standing at the threshold of a product management renaissance powered by artificial intelligence at full scale.

The coming years will be defined by rapid capability gains and equally rapid shifts in expectations. PMs will soon wield tools once reserved for data scientists and futurists: real-time foresight, limitless ideation partners, and analytics engines that make complexity feel simple. This power is thrilling, but it carries responsibility. You will be expected to navigate ethical complexity, anticipate market shifts before they hit, and lead teams through constant reinvention.

This chapter offers a forward-looking roadmap. It spotlights emerging developments, looming disruptions, and the skills and strategies that will set tomorrow's leaders apart.

Riding the Next Wave

The next wave of AI isn't abstract. It is already reshaping product management. What feels like an edge today will be table stakes tomorrow. The PMs who adapt early won't just keep up, they'll set the pace.

AI Democratization and Accessibility
AI will soon be as standard as spreadsheets. No-code platforms will let PMs without technical backgrounds integrate advanced features directly into workflows. The skill shift: every PM must learn how to frame problems for AI, interpret outputs, and fold results into the product lifecycle with confidence.

Human-AI Collaboration
AI is moving from silent assistant to active collaborator, co-creating product visions, running live scenarios, and facilitating stakeholder conversations in real

conversations in real time. PMs will need adaptive leadership to guide these conversations, keeping focus on strategy while AI handles the complexity.

Real-Time Predictive Product Intelligence

Predictive analytics won't live in quarterly decks anymore. They'll appear on dashboards updated by the hour. A PM might see churn signals on Friday, model three responses before lunch, and ship fixes by Monday. The skill shift: scenario-based strategic thinking and comfort with rapid iteration.

Ethical AI as Core Practice

Trust is fragile but non-negotiable. PMs will own fairness, accountability, and compliance as part of their role. That requires ethical judgment, data stewardship, and the courage to treat ethics as both a design constraint and a competitive advantage.

Cross-Disciplinary Fluency

The strongest PMs will blend AI literacy with behavioral science, ethics, and classic product management principles. No single discipline is enough on its own; the edge comes from weaving them together into decisions that are both data-sharp and human-grounded.

AI Product Management Horizon Map
3–5 Year Outlook

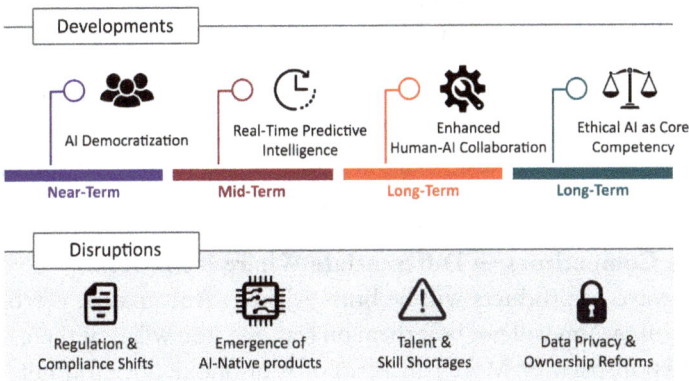

Developments			
AI Democratization	Real-Time Predictive Intelligence	Enhanced Human-AI Collaboration	Ethical AI as Core Competency
Near-Term	**Mid-Term**	**Long-Term**	**Long-Term**

Disruptions			
Regulation & Compliance Shifts	Emergence of AI-Native products	Talent & Skill Shortages	Data Privacy & Ownership Reforms

Leading at MACH-10 means pairing speed with soul

These developments represent massive opportunity, but only for PMs who see them coming and adjust ahead of the curve. The gap won't be between companies that use AI and those that do not, but between leaders who weave it seamlessly into strategy and those who treat it as an add-on.

This is where skillset becomes the deciding factor. The ability to harness AI's promise while steering clear of its pitfalls will define the next generation of product leaders.

> **" The future belongs to product leaders who embrace AI not as a threat, but as the greatest force multiplier for human ingenuity "**

The Headwinds You'll Face

Every leap forward brings friction. These disruptions are already here. How you respond will decide whether you adapt or get left behind.

Regulation and Compliance Shifts → Stay Ahead of the Curve
New rules are coming fast. Think GDPR 2.0 with sharper teeth. If you wait until laws hit, you have already lost. Build flexible frameworks now so you can bend without breaking. Resilience is the new compliance.

AI-Native Competitors → Differentiate Where It Matters
The next wave of products will be born AI-first. Automated. Personalized. Always learning. You will not beat them on features. You will beat them on trust, clarity, and transparency. Make ethics part of your brand, not a footnote.

Talent and Skill Gaps → Don't Chase, Build
The hunt for AI-fluent PMs is already fierce. Stop waiting for unicorn hires. Train your team now. Drip learning into the culture every week and let compound knowledge do the work. By the time everyone else is scrambling for talent, you will already have it.

Data Ownership and Privacy Reforms → Lead with Stewardship
Data control is shifting back to users. That will reshape research, personalization, and monetization. Teams that see this as a burden will stall. Teams that lean into stewardship will build trust, loyalty, and a long-term moat.

Here is the truth: disruption does not decide the future. Response does. The PMs who thrive are not just adopting tools. They are building systems, habits, and cultures that turn volatility into advantage.

At the center of this is the PM. Your role is not to collect tools but to set the tone for how AI is understood, integrated, and trusted. By investing in education, frameworks, ethics, and foresight, you shift from reacting to shaping. That is the difference between keeping up and leading.

These principles are the foundation. But strategy only becomes real when it meets execution. The fastest way to start is by experimenting with tools that bring these ideas to life. The next section highlights five practical moves you can test this week to build momentum.

Don't Wait. Define.

The future of product management will not be defined by who adopts AI tools. It will be defined by who learns to lead with them. The difference between keeping up and setting the pace is simple: foresight, courage, and the willingness to reinvent before you are forced to.

AI is your multiplier, but leadership is your edge. Machines can analyze, simulate, and optimize. They cannot rally a team through uncertainty, shape a vision people believe in, or carry the responsibility for what gets built and why. That remains your work.

The PMs who thrive in this future will not just react to disruption. They will orchestrate it. They will treat ethics as strategy, foresight as discipline, and AI as both lens and lever.

You do not need to predict the decade ahead. You need to prepare to shape it. The choice is here now: wait and adapt, or lead and define. The MACH-10 PM chooses the second path.

The next era of PMs will not just manage products. They will multiply possibility.

5 Tools to Try This Week

OpenAI Playground	Experiment with advanced AI prompt scenarios for strategic foresight
Notion	Document evolving AI capabilities and team learning plans
miro	Visually map future org structures and workflows incorporating AI
ETHICS GRADE	Assess ethical compliance of AI initiatives
Airtable	Track regulatory changes and compliance readiness

3 Prompts to Test Today

1. List emerging AI trends that will impact product management over the next 3 years.

2. Design a roadmap for building AI fluency within a product team.

3. Identify potential ethical risks and mitigation strategies for AI product development.

Mindset Shift to Adopt

The future belongs to PMs who see AI not as a threat but as a multiplier of human ingenuity. This is not about replacing human capability. It is about amplifying it. Your creativity, empathy, and judgment remain irreplaceable. AI sharpens them.

Reflective Questions

Which AI developments are most likely to reshape your role, your team, or your industry in the next few years?

How ready is your organization for changes in AI regulations, ethical standards, or market dynamics that could hit faster than expected?

Where are the biggest gaps between your current capabilities and the skills that will be required to thrive?

What bold step could you take in the next 90 days to position yourself and your team for the hypersonic future?

Exercise

AI is reshaping product organizations as radically as the shift from waterfall to agile. Teams will be smaller, faster, and more adaptive. Human creativity and machine efficiency will work as one, freeing people to focus on innovation and strategy.

The teams that win will test, learn, and ship at speeds we can barely imagine. New roles will emerge: AI Systems Architects, Intelligence Orchestrators, embedded copilots, and staffing models that fill gaps instantly.

The future will reward leaders who orchestrate outcomes with precision and vision. They will build trust, inspire teams, and turn human and AI collaboration into a force for growth.

Key Takeaways

- AI is your force multiplier. It compounds clarity and compresses time, but only if you steer it.

- **Disruptions are guaranteed.** Regulations, AI-native competitors, talent gaps, data reforms. You can't dodge them, but you can outpace them.

- **Your advantage isn't tools, it's leadership.** Machines can analyze and simulate. Only you can rally a team, set a vision, and own the "why."

- **The winning PM is built, not found.** Keep investing in AI fluency, ethical judgment, and adaptive leadership. That's the compound interest that separates leaders from laggards.

- **Adaptability is the culture.** Make resilience and responsibility your operating system. The orgs that do this won't just survive the AI wave — they'll shape it.

Final Word: Embrace the MACH-10 PM Mindset

Lead at MACH-10

When I first stepped into product management, the landscape was unrecognizable. Google was a curiosity, Jira and Confluence didn't exist, and Agile was more idea than standard practice. At Qualcomm, long cycles and global teams left no room for iteration. At GoPro, I saw how fast markets can outrun even the best systems. Every PM learns this truth: the role never stands still.

Science fiction warned of an AI apocalypse. *Blade Runner* showed us soulless replicants. *The Terminator* promised machines that would rise and erase us.

Reality delivered the opposite: AI is making us more human. It doesn't replace judgment. It magnifies it, cutting through noise so we can focus on strategy, empathy, and vision. The best product leaders keep evolving with their tools, their markets, and the expectations that keep climbing.

AI is reshaping the product journey: faster discovery, forward-looking roadmaps, adaptive delivery, and leadership grounded in ethics and experimentation. These shifts compound. Focus fuels speed. Speed fuels insight. Being a MACH-10 PM isn't about chasing tools. It's about leverage. Combine AI with judgment. Build systems where insight drives action. Set a higher bar for conviction. Power at MACH-10 demands integrity at MACH-10.

Looking ahead, the most impactful PMs won't be judged by team size or feature count. They'll be defined by how well they orchestrate human and AI intelligence into outcomes competitors can't match. Those who master this shift will set the pace. Those who don't will fall behind.

No one will remember every feature you shipped. They'll remember the clarity you brought in the fog, the courage you showed in uncertainty, and the people you lifted along the way. That's the legacy: speed with soul. This book is a launchpad. Lead smarter, not just faster. Lead more meaningfully, not just more efficiently.

Tomorrow morning, before you open Slack or dive into Jira, pick one product decision. Run it through an AI tool with a prompt from this book. Compare it to your instinct. Notice the difference. That's the start of your MACH-10 mindset. To keep your edge sharp, visit mach10pm.com for new playbooks, case studies, and prompt libraries: the extended manual for sustaining speed with soul.

Do not predict the future. Shape it.

The MACH-10 PM isn't just a role. It's a responsibility to move fast, think deeply, and lead with purpose. You have the frameworks and the mindset. Go multiply the impact.

"Being a MACH-10 PM means mastering the art of speed with soul: leading with purpose, precision, and AI-augmented insight."
-Jason M. Riggs

Build fast. Build people. Leave a legacy.